The Best American Science Writing 2008

THE BEST AMERICAN SCIENCE WRITING

The Best American

2008

SCIENCE WRITING

EDITOR: SYLVIA NASAR

Series Editor: Jesse Cohen

AN ecco BOOK

HARPER PERENNIAL

NEW YORK • LONDON • TORONTO • SYDNEY • NEW DELHI • AUCKLAND

HARPER ● PERENNIAL

FIRST EDITION

Library of Congress Cataloging-in-Publication Data is available upon request.

ISBN 978-0-06-134041-3

08 09 10 11 12 WBC/RRD 10 9 8 7 6 5 4 3 2 1

Contents

Introduction by Sylvia Nasar

LIKE MARKETS AND DEMOCRACY, science is doing an incredible job of proliferating choice. From instant communication to genetic testing, fertility treatments to space tourism, alternative fuels to genetically modified corn, the options open to the average American are dizzying, exhilarating, kaleidoscopic, and unprecedented. As Pandora, Faust, and others warned, plethoric choice has its drawbacks. It's confusing, arduous, burdensome, worrying, risky, and conflict-provoking—the very things, I'd like to point out, that make people want to turn to science journalism or feel compelled to practice it. All that uncertainty and anxiety and conflict are like the sand in an oyster's shell: they inspire pearls of reporting and writing.

No entry in this volume illustrates the point better than Amy Harmon's haunting story about truth and its consequences. I kept being reminded of her piece while in New York City, attending the Pearl Theatre's production of Henrik Ibsen's *Ghosts*. Like Ibsen's Victorian tragedy, Harmon's story, whose subject is ostensibly the impact on society of breakthroughs in genetic testing, is really a drama about family secrets, inherited disease, diabolical punishments inflicted on the innocent, and the primal conflict between mother and child. The story, for which Ms. Harmon has since won a Pulitzer, concerns a young woman whose grandfather died of Huntington's chorea, a rare but cruel disease caused by a single gene. Huntington's is invariably deadly but death comes with agonizing slowness. The first debilitating symptoms typically appear in one's thirties or forties and most victims die within a decade or so.

Having one grandparent with the disease places the average odds at one in two, but the actual odds, if she in fact carries the lethal gene, are one in one or, if she does not, zero. Which is it, the lady or the tiger? Ignoring her mother's warnings, the young woman insists on finding out the truth. The result is the one that she feared. She does, in fact, have the lethal gene. Other tests show that she has twelve years before her power to "walk, talk, and think" start to go.

Harmon's story is about the repercussions of the woman's fateful choice to know. One is an escalating conflict with her estranged mother that results in the story's explosive climax. The mother doesn't know that the daughter actually got tested and she herself has chosen not to be tested because she would rather not know. "You want to enjoy life," she explains to Harmon.

The dramatic denouement occurs during a custody hearing over the young woman's cousin, and the young woman decides to testify against her mother. She tells the judge that she has tested positive for the lethal Huntington gene. One glance at her mother's "stricken" expression shows that the mother had grasped the fact of her death sentence. The daughter could only carry the gene if she had inherited it from her mother. This is the kind of scene that makes a theater audience gasp.

I KNEW I WANTED TO include Ms. Harmon's story in our anthology because I haven't stopped thinking about it. Which brings me to another kind of choice, namely my choice of stories in this volume, which were, I hasten to add, hugely helped by my astute and always amiable partner on this project, series editor Jesse Cohen, and the suggestions of two discerning friends who are science writers, neurologist Orly Avitzur and marine biologist David Gruber.

I'm happy to report that choosing stories for this volume involved a great deal more pleasure than agony. The topics pretty much reflect what was in the news and on people's minds last year. Math, which made headlines in 2006 with the Poincaré conjecture, disappeared from sight in 2007. Ditto physics. Instead, the environment and health monopolized attention. In 2007, all eyes were on melting ice caps, energy, Al Gore's Oscar and Nobel, genetic tests, and tainted products from China.

I gravitated to stories that people talked about, that stuck in my mind long after I read them, or that gave the received wisdom a jolt big enough to shift public opinion on some important issue. But while the choices certainly reflect my personal tastes and interests, I think that you will find that all of them share certain qualities. Three attributes make these stories not just great science writing but great journalism: a compelling story idea, not just a topic; extraordinary, often exclusive reporting; and a facility for concisely expressing complex ideas and masses of information.

Richard Preston's "An Error in the Code" is impossible to put down. It's a puzzle, a quest, and a human dilemma, but in an extreme, larger-than-life way. To be attacked by your own hands is incredibly bizarre. To gouge out your eyes or bite off the tips of your fingers is grotesque and terrifying. Yet Mr. Preston depicts the men and boys in the story as loving, funny, self-aware individuals. He evokes terror and pity, but also humor and compassion.

Don't let Mr. Preston's virtuosity as a reporter and storyteller blind you, however, to the consummate skill and authority with which he conveys the significance and substance of the science involved. He does the former by contrasting the obscurity of the disease with the universality of some of its features: Lesch-Nyhan syndrome is so rare that one of the eponymous medical researchers knows virtually every individual afflicted with it, yet the condition may serve as a template for "trying to trace the action of one gene on complex human behavior." Mr. Preston's description of the link between gene mutations and behavior is a jewel of a science lesson: lucid, self-contained, and concise, yet studded with examples and illustrations.

Wall Street Journal reporter Tara Parker-Pope's job is to help consumers make smarter choices about health care. Not exactly rocket science, right? Wrong. Her story about the NIH's misinterpretation of a high-profile study of the heart risks of hormone replacement therapy is another one that I remembered long after I read it. Ms. Parker-Pope does something all too few reporters think to do: revisits a major news story from a few years back to ask whether the scientists and the media got it right. Her analysis of this 2002 study, which scared millions of low-risk women into abandoning hormone replacement, is a model of thorough reporting and nuanced yet clear writing. I included her guide for women because it's hard to do what she does—actually spelling out what actions are consistent with the study's conclusion—well.

CONFLICTS

Choice creates conflicts, both internal and among individuals and groups, and conflicts often influence and complicate individual choices. This is doubly true of hidden conflicts of interest. If last year's crop of investigative science stories is any indication, such conflicts seem to be especially rampant in psychiatry.

Familiar with bestsellers like *Covert: My Years Infiltrating the Mob*? *The Insider*? *Confessions of an Economic Hit Man*? First-person journalism by turncoats who rat out their fellows is not only eye-opening; it takes us places it would be hard for other journalists to go and reveals all kinds of delicious details. Boston psychiatrist Dan Carlat is that oxymoron, a self-employed whistle-blower. He describes a year spent promoting Effexor, an antidepressant made by Wyeth, to fellow psychiatrists as well as internists and others who are less than expert on the efficacy and risks of psychiatric drugs. Incredibly, he betrayed his professional ethics for a pittance: $750 per speech. He claims that one quarter of M.D.s do it, though he doesn't report what fraction of psychiatrists do. What would the American Medical Association say if the *New York Times*'s tax reporter decided to moonlight on H&R Block's PR team? Given that M.D.s live and die by referrals, Dr. Carlat's decision to expose dubious practices in his trade is either pretty brave or else a sign that he thinks there's more money to be made in publishing.

Few patients are more vulnerable than the children and their often desperate parents who turn up in psychiatrists' offices. That is what makes the investigative story by Gardiner Harris, Benedict Carey, and Janet Roberts about psychiatrists who collect fees to market certain antipsychotic drugs while simultaneously prescribing these same drugs to their child patients such a blockbuster. The *New York Times* reporters took advantage of the fact that Minnesota is the only state to make public data on drug company payments to doctors and on prescribing patterns. Harris and Roberts scored a second coup by comparing two seemingly unrelated databases also compiled by Minnesota: physicians paid by pharmaceutical companies to supervise drug trials versus doctors who had been sanctioned by the state's medical board. The mere idea of lining the two lists up suggests a suspicious, if not downright paranoid, cast of mind that is essential for successful investigative journalism. After all, clinical investigators are supposed to be recruited from the crème de la crème of the profession, aren't they? As it turns out, paranoia was perfectly justified. Drug companies hired more than a hundred Minnesota doctors with dirty records in the last decade, including a former president of the Minnesota Psychiatric Society whose license was suspended for reckless disregard for the welfare of forty-six patients, five of whom died.

A more subtle conflict—between the vow of doctors to do no harm and their desire to provide patients with answers—is at the heart of one of last year's timeliest and most talked about medical stories. Focusing on the diagnosis du jour in child psychiatry, pediatric bipolar disorder, *The New Yorker's* Jerome Groopman casts a skeptical eye on one of the most common pitfalls in a fashion-conscious, media-driven society. Thanks to a bestselling book, a breathless *Time* magazine cover story, and a handful of zealous doctors, legions of parents are now convinced that their children are suffering from bipolar disorder, until recently thought not to occur before adolescence. Dr. Groopman reports that children as young as two years old are being treated with powerful antipsychotic drugs. Not only are the diagnostic criteria elastic, but the risks are poorly understood given that none of these drugs were ever tested on children, much less tailored to them. Without sounding either dismissive or alarmist, Groopman marshals data, case studies, and clinical experience to make a quietly compelling case against any parent or doctor climbing on this particular bandwagon.

THE BRAIN

Comparing last year's science stories to the science stories that people were reading a hundred or a hundred fifty years ago, I'm struck that the brain is back in the limelight just as it was during Queen Victoria's reign, when many thinking people considered themselves disciples of Ernst Mach, a brilliant Austrian physicist who turned from the stars to the senses. For many reasons, not least the plummeting price of computing and rapid advances in molecular biology, the brain is becoming more of an open book and there is more new stuff to report even on old problems like lying.

Some of the best writing about the brain last year was the most personal. Margaret Talbot's story about the shaky science of telling truth from falsehood is delicious partly because she begins with a spectacular instance of being duped herself and then proceeds to demonstrate, throughout the rest of her story, an astringent and welcome skepticism toward the overconfident claims of some of her subjects. In "The Abyss," about a man condemned to live, literally, in the present, Oliver Sacks reminds me of Trollope, whose novels were full of odd, lonely, suffering people he encountered in his professional life—setting up postal routes in rural England as opposed to treating patients, as Dr. Sacks does. Trollope and Dr. Sacks share that wonderful English ability to sympathize without sentimentalizing.

The Globe

The *biggest* story last year, without a doubt, was the planet itself. In January 2007, a group of government scientists employed at NASA and elsewhere reported that the polar ice caps were melting much faster than previously thought. A few weeks later, *An Inconvenient Truth*, Al Gore's documentary about global warming, won an Oscar. In June came the revelation that the White House was editing the testimony of senior scientists to downplay the threat of climate change. Finally, in October, the former vice president, presidential candidate, and journalist won the Nobel Peace Prize. Result: global warming became *the* science story.

I include a short piece by Vice President Gore mainly to underscore that defining the story can be as powerful as telling it. The right story idea at the right time becomes a defining moment in a war or a presidency or an issue.

Among the most interesting environmental stories is the dilemma posed by China's economic miracle. Two of the freshest pieces were part of a *New York Times* series on pollution in China. Jim Yardley's "Beneath Booming Cities, China's Future Is Drying Up," is authoritatively reported and strongly told. I also found Joseph Kahn's story about the environmental activist very moving.

The last selection in the volume is John Seabrook's story about a global seed bank. In March 2008, when a big chunk of Antarctic ice broke off and floated out to sea, it is somehow comforting to think that there is a sort of agricultural Noah's Ark being launched in Norway and that, in fact, there have been and are and will be many such banks scattered around the world. Many of the stories Mr. Seabrook tells are heartbreaking: the random destruction of the Iraqi seed bank in the U.S.-led invasion, the deliberate destruction by Stalin of the Soviet seed specialists, the decrepit state of the current Russian facility. But the very subject of seeds and the lengths to which people will go to pass them on is like spring, hopeful.

Sylvia Nasar
March 27, 2008
Graduate School of Journalism
Columbia University
New York City

The Best American Science Writing 2008

Amy Harmon

Facing Life with a Lethal Gene

FROM THE *NEW YORK TIMES*

Now that genetic testing is becoming more common people at risk for hereditary disease are struggling with whether to find out their genetic fates. Amy Harmon follows one young woman whose test results have changed her life.

THE TEST, THE COUNSELOR SAID, had come back positive.

Katharine Moser inhaled sharply. She thought she was as ready as anyone could be to face her genetic destiny. She had attended a genetic counseling session and visited a psychiatrist, as required by the clinic. She had undergone the recommended neurological exam. And yet, she realized in that moment, she had never expected to hear those words.

"What do I do now?" Ms. Moser asked.

"What do you want to do?" the counselor replied.

"Cry," she said quietly.

Her best friend, Colleen Elio, seated next to her, had already begun.

Ms. Moser was twenty-three. It had taken her months to convince the clinic at New York-Presbyterian Hospital/Columbia University Medical Center in Manhattan that she wanted, at such a young age, to find out whether she carried the gene for Huntington's disease.

Huntington's, the incurable brain disorder that possessed her grandfather's body and ravaged his mind for three decades, typically strikes in middle age. But most young adults who know the disease runs in their family have avoided the DNA test that can tell whether they will get it, preferring the torture—and hope—of not knowing.

Ms. Moser is part of a vanguard of people at risk for Huntington's who are choosing to learn early what their future holds. Facing their genetic heritage, they say, will help them decide how to live their lives.

Yet even as a raft of new DNA tests are revealing predispositions to all kinds of conditions, including breast cancer, depression, and dementia, little is known about what it is like to live with such knowledge.

"What runs in your own family, and would you want to know?" said Nancy Wexler, a neuropsychologist at Columbia and the president of the Hereditary Disease Foundation, which has pioneered Huntington's research. "Soon everyone is going to have an option like this. You make the decision to test, you have to live with the consequences."

On that drizzly spring morning two years ago, Ms. Moser was feeling her way, with perhaps the most definitive and disturbing verdict genetic testing has to offer. Anyone who carries the gene will inevitably develop Huntington's.

She fought her tears. She tried for humor.

Don't let yourself get too thin, said the clinic's social worker. Not a problem, Ms. Moser responded, gesturing to her curvy frame. No more than two drinks at a time. Perhaps, Ms. Moser suggested to Ms. Elio, she meant one in each hand.

Then came anger.

"Why me?" she remembers thinking, in a refrain she found hard to shake in the coming months. "I'm the good one. It's not like I'm sick because I have emphysema from smoking or I did something dangerous."

The gene that will kill Ms. Moser sits on the short arm of everyone's fourth chromosome, where the letters of the genetic alphabet normally repeat C-A-G as many as thirty-five times in a row. In people who develop Huntington's, however, there are more than thirty-five repeats.

No one quite knows why this DNA hiccup causes cell death in the brain, leading Huntington's patients to jerk and twitch uncontrollably and rendering them progressively unable to walk, talk, think, and swallow. But the greater the number of repeats, the earlier symptoms tend to appear and the faster they progress.

Ms. Moser's "CAG number" was forty-five, the counselor said. She had more repeats than her grandfather, whose first symptoms—loss of short-term memory, mood swings, and a constant ticking noise he made with his mouth—surfaced when he turned fifty. But it was another year before Ms. Moser would realize that she could have less than twelve years until she showed symptoms.

Immediately after getting her results, Ms. Moser was too busy making plans.

"I'm going to become super-strong and super-balanced," she vowed over lunch with Ms. Elio, her straight brown hair pulled into a determined bun. "So when I start to lose it I'll be a little closer to normal."

In the tumultuous months that followed, Ms. Moser often found herself unable to remember what normal had once been. She forced herself to renounce the crush she had long nursed on a certain firefighter, sure that marriage was no longer an option for her. She threw herself into fund-raising in the hopes that someone would find a cure. Sometimes, she raged.

She never, she said, regretted being tested. But at night, crying herself to sleep in the dark of her lavender bedroom, she would go over and over it. She was the same, but she was also different. And there was nothing she could do.

LESSON IN STIGMA

Ms. Moser grew up in Connecticut, part of a large Irish Catholic family. Like many families affected by Huntington's, Ms. Moser's regarded the disease as a curse, not to be mentioned even as it dominated their lives in the form of her grandfather's writhing body and unpredictable rages.

Once, staying in Ms. Moser's room on a visit, he broke her trundle bed with his violent, involuntary jerking. Another time, he came into the kitchen naked, his underpants on his head. When the children giggled, Ms. Moser's mother defended her father: "If you don't like it, get out of my house and go."

But no one explained what had happened to their grandfather, Thomas Dowd, a former New York City police officer who once had dreams of retiring to Florida.

In 1990, Mr. Dowd's older brother, living in a veteran's hospital in an advanced stage of the disease, was strangled in his own restraints. But a year or so later, when Ms. Moser wanted to do her sixth-grade science project on Huntington's, her mother recoiled.

"Why," she demanded, "would you want to do it on this disease that is killing your grandfather?"

Ms. Moser was left to confirm for herself, through library books and a CD-ROM encyclopedia, that she and her brothers, her mother, her aunts, an uncle, and cousins could all face the same fate.

Any child who has a parent with Huntington's has a 50 percent chance of having inherited the gene that causes it, Ms. Moser learned.

Her mother, who asked not to be identified by name for fear of discrimination, had not always been so guarded. At one point, she drove around with a "Cure HD" sign in the window of her van. She told people that her father had "Woody Guthrie's disease," invoking the folk icon who died of Huntington's in 1967.

But her efforts to raise awareness soon foundered. Huntington's is a rare genetic disease, affecting about 30,000 people in the United States, with about 250,000 more at risk. Few people know what it is. Strangers assumed her father's unsteady walk, a frequent early symptom, meant he was drunk.

"Nobody has compassion," Ms. Moser's mother concluded. "People look at you like you're strange, and 'What's wrong with you?'"

Shortly after a simple DNA test became available for Huntington's in 1993, one of Ms. Moser's aunts tested positive. Another, driven to find out if her own medical problems were related to Huntington's, tested negative. But when Ms. Moser announced as a teenager that she wanted to get tested one day, her mother insisted that she should not. If her daughter carried the gene, that meant she did, too. And she did not want to know.

"You don't want to know stuff like that," Ms. Moser's mother said in an interview. "You want to enjoy life."

Ms. Moser's father, who met and married his wife six years before Ms. Moser's grandfather received his Huntington's diagnosis, said he had managed not to think much about her at-risk status.

"So she was at risk," he said. "Everyone's at risk for everything."

The test, Ms. Moser remembers her mother suggesting, would cost thousands of dollars. Still, in college, Ms. Moser often trolled the Web for information about it. Mostly, she imagined how sweet it would be to know she did not have the gene. But increasingly she was haunted, too, by the suspicion that her mother did.

As awful as it was, she admitted to Ms. Elio, her freshman-year neighbor at Elizabethtown College in Pennsylvania, she almost hoped it was true. It would explain her mother's strokes of meanness, her unpredictable flashes of anger.

Ms. Moser's mother said she had never considered the conflicts with her daughter out of the ordinary. "All my friends who had daughters said that was all normal, and when she's twenty-five she'll be your best friend," she said. "I was waiting for that to happen, but I guess it's not happening."

When Ms. Moser graduated in 2003 with a degree in occupational therapy, their relationship, never peaceful, was getting worse. She moved to Queens without giving her mother her new address.

WANTING TO KNOW

Out of school, Ms. Moser soon spotted a listing for a job at Terence Cardinal Cooke Health Care Center, a nursing home on the Upper East Side of Manhattan. She knew it was meant for her.

Her grandfather had died there in 2002 after living for a decade at the home, one of only a handful in the country with a unit devoted entirely to Huntington's.

"I hated visiting him growing up," Ms. Moser said. "It was scary."

Now, though, she was drawn to see the disease up close.

On breaks from her duties elsewhere, she visited her cousin James Dowd, the son of her grandfather's brother who had come to live in the Huntington's unit several years earlier. It was there, in a conversation with another staff member, that she learned she could be tested for only a few hundred dollars at the Columbia clinic across town. She scheduled an appointment for the next week.

The staff at Columbia urged Ms. Moser to consider the downside of genetic testing. Some people battle depression after they test positive. And the information, she was cautioned, could make it harder for her to get a job or health insurance.

But Ms. Moser bristled at the idea that she should have to remain ignorant about her genetic status to avoid discrimination. "I didn't do anything wrong," she said. "It's not like telling people I'm a drug addict."

She also recalls rejecting a counselor's suggestion that she might have asked to be tested as a way of crying for help.

"I'm like, 'No,'" Ms. Moser recalls replying. "'I've come to be tested because I want to know.'"

No one routinely collects demographic information about who gets tested for Huntington's. At the Huntington's Disease Center at Columbia, staff members say they have seen few young people taking the test.

Ms. Moser is still part of a distinct minority. But some researchers say her attitude is increasingly common among young people who know they may develop Huntington's.

More informed about the genetics of the disease than any previous generation, they are convinced that they would rather know how many healthy years they have left than wake up one day to find the illness upon them. They are confident that new reproductive technologies can allow them to have children without transmitting the disease and are eager to be first in line should a treatment become available.

"We're seeing a shift," said Dr. Michael Hayden, a professor of human genetics at the University of British Columbia in Vancouver who has been providing various tests for Huntington's for twenty years. "Younger people are coming for testing now, people in their twenties and early thirties; before, that was very rare. I've counseled some of them. They feel it is part of their heritage and that it is possible to lead a life that's not defined by this gene."

Before the test, Ms. Moser made two lists of life goals. Under "if negative," she wrote *married*, *children*, and *Ireland*. Under "if positive" was *exercise*, *vitamins*, and *ballroom dancing*. Balance, in that case, would be important. Opening a bed-and-breakfast, a goal since childhood, made both lists.

In the weeks before getting the test results, Ms. Moser gave Ms. Elio explicit instructions about acceptable responses. If she was negative, flowers were OK. If positive, they were not. In either case, drinking was acceptable. Crying was not.

But it was Ms. Elio's husband, Chris Elio, who first broached the subject of taking care of Ms. Moser, whom their young children called "my Katie," as in, "this is my mom, this is my dad, this is my Katie." They should address it before the results were in, Mr. Elio told his wife, so that she would not feel, later, that they had done it out of a sense of obligation.

The next day, in an e-mail note that was unusually formal for friends who sent text messages constantly and watched *Desperate Housewives* while on the phone together, Ms. Elio told Ms. Moser that she and her husband wanted her to move in with them if she got sick. Ms. Moser set the note aside. She did not expect to need it.

"IT'S TOO HARD TO LOOK"

The results had come a week early, and Ms. Moser assured her friends that the *Sex and the City* trivia party she had planned for that night was still on. After all, she was not sick, not dying. And she had already made the dips.

"I'm the same person I've always been," she insisted that night as her guests gamely dipped strawberries in her chocolate fountain. "It's been in me from the beginning."

But when she went to work the next day, she lingered outside the door of the occupational therapy gym, not wanting to face her colleagues. She avoided the Huntington's floor entirely, choosing to attend to patients ailing from just about anything else. "It's too hard to look at them," she told her friends.

In those first months, Ms. Moser summoned all her strength to pretend that nothing cataclysmic had happened. At times, it seemed easy enough. In the mirror, the same green eyes looked back at her. She was still tall, a devoted Julia Roberts fan, a prolific baker.

She dropped the news of her genetic status into some conversations like small talk, but kept it from her family. She made light of her newfound fate, though often friends were not sure how to take the jokes.

"That's my Huntington's kicking in," she told Rachel Markan, a coworker, after knocking a patient's folder on the floor.

Other times, Ms. Moser abruptly dropped any pretense of routine banter. On a trip to Florida, she and Ms. Elio saw a man in a wheelchair being tube-fed, a method often used to keep Huntington's patients alive for years after they can no longer swallow.

"I don't want a feeding tube," she announced flatly.

In those early days, she calculated that she had at least until fifty before symptoms set in. That was enough time to open a bed-and-breakfast, if she acted fast. Enough time to repay seventy thousand dollars in student loans under her thirty-year term.

Doing the math on the loans, though, could send her into a tailspin.

"I'll be repaying them and then I'll start getting sick," she said. "I mean, there's no time in there."

FINDING NEW PURPOSE

At the end of the summer, as the weather grew colder, Ms. Moser forced herself to return to the Huntington's unit.

In each patient, she saw her future: the biophysicist slumped in his wheelchair, the refrigerator repairman inert in his bed, the onetime professional tennis player who floated through the common room, arms undulating in the startlingly graceful movements that had earned the disease its original name, "Huntington's chorea," from the Greek "to dance."

Then there was her cousin Jimmy, who had wrapped papers for the *New York Post* for nineteen years until suddenly he could no longer tie the knots. When she greeted him, his bright blue eyes darted to her face, then away. If he knew her, it was impossible to tell.

She did what she could for them. She customized their wheelchairs with padding to fit each one's unique tics. She doled out special silverware, oversized or bent in just the right angles to prolong their ability to feed themselves.

Fending off despair, Ms. Moser was also filled with new purpose. Someone, somewhere, she told friends, had to find a cure.

It has been over a century since the disease was identified by George Huntington, a doctor in Amagansett, New York, and over a decade since researchers first found the gene responsible for it.

To raise money for research, Ms. Moser volunteered for walks and dinners and golf outings sponsored by the Huntington's Disease Society of America. She organized a Hula-Hoop-a-thon on the roof of Cardinal Cooke, then a bowl-a-thon at the Port Authority. But at many of the events, attendance was sparse.

It is hard to get people to turn out for Huntington's benefits, she learned from the society's professional fund-raisers. Even families affected by the disease, the most obvious constituents, often will not help publicize events.

"They don't want people to know they're connected to Huntington's," Ms. Moser said, with a mix of anger and recognition. "It's like in my family—it's not a good thing."

Her first session with a therapist brought a chilling glimpse of how the disorder is viewed even by some who know plenty about it. "She told me it was my moral and ethical obligation not to have children," Ms. Moser told Ms. Elio by cellphone as soon as she left the office, her voice breaking.

In lulls between fund-raisers, Ms. Moser raced to educate her own world about Huntington's. She added links about the disease to her MySpace page. She plastered her desk at work with "Cure HD" stickers and starred in a video about the Huntington's unit for her union's Web site.

Ms. Moser gave blood for one study and spoke into a microphone for researchers trying to detect subtle speech differences in people who have extra CAG repeats before more noticeable disease symptoms emerge.

When researchers found a way to cure mice bred to replicate features of the disease in humans, Ms. Moser sent the news to friends and acquaintances.

But it was hard to celebrate. "Thank God," the joke went around on the Huntington's National Youth Alliance e-mail list Ms. Moser subscribed to, "at least there won't be any more poor mice wandering around with Huntington's disease."

In October, one of Ms. Moser's aunts lost her balance while walking and broke her nose. It was the latest in a series of falls. "The cure needs to be soon for me," Ms. Moser said. "Sooner for everybody else."

A Confrontation in Court

In the waiting room of the Dutchess County family courthouse on a crisp morning in the fall of 2005, Ms. Moser approached her mother, who turned away.

"I need to tell her something important," Ms. Moser told a family member who had accompanied her mother to the hearing.

He conveyed the message and brought one in return: unless she was dying, her mother did not have anything to say to her.

That Ms. Moser had tested positive meant that her mother would develop Huntington's, if she had not already. A year earlier, Ms. Moser's mother had convinced a judge that her sister, Nora Maldonado, was neglecting her daughter. She was given guardianship of the daughter, four-year-old Jillian.

Ms. Moser had been skeptical of her mother's accusations that Ms. Maldonado was not feeding or bathing Jillian properly, and she wondered whether her effort to claim Jillian had been induced by the psychological symptoms of the disease.

Her testimony about her mother's genetic status, Ms. Moser knew, could help persuade the judge to return Jillian. Ms. Maldonado had found out years earlier that she did not have the Huntington's gene.

Ms. Moser did not believe that someone in the early stages of Huntington's should automatically be disqualified from taking care of a child. But her own rocky childhood had convinced her that Jillian would be better off with Ms. Maldonado.

She told her aunt's lawyer about her test results and agreed to testify.

In the courtroom, Ms. Moser took the witness stand. Her mother's lawyer jumped up as soon as the topic of Huntington's arose. It was irrelevant, he said. But by the time the judge had sustained his objections, Ms. Moser's mother, stricken, had understood.

The next day, in the bathroom, Ms. Maldonado approached Ms. Moser's mother.

"I'm sorry," she said. Ms. Moser's mother said nothing.

The court has continued to let Ms. Moser's mother retain guardianship of Jillian. But she has not spoken to her daughter again.

"It's a horrible illness," Ms. Moser's mother said, months later, gesturing to her husband. "Now he has a wife who has it. Did she think of him? Did she think of me? Who's going to marry her?"

FACING THE FUTURE

Before the test, it was as if Ms. Moser had been balanced between parallel universes, one in which she would never get the disease and one in which she would. The test had made her whole.

She began to prepare the Elio children and Jillian for her illness, determined that they would not be scared, as she had been with her grandfather. When Jillian wanted to know how people got Huntington's disease "in their pants," Ms. Moser wrote the text of a children's book that explained what these other kinds of "genes" were and why they would make her sick.

But over the winter, Ms. Elio complained gently that her friend had become "Ms. H.D." And an impromptu note that arrived for the children in the early spring convinced her that Ms. Moser was dwelling too much on her own death.

"You all make me so happy, and I am so proud of who you are and who you will be," read the note, on rainbow scratch-and-write paper. "I will always remember the fun things we do together."

Taking matters into her own hands, Ms. Elio created a profile for

Ms. Moser on an online dating service. Ms. Moser was skeptical but supplied a picture. Dating, she said, was the worst thing about knowing she had the Huntington's gene. It was hard to imagine someone falling enough in love with her to take on Huntington's knowingly, or asking it of someone she loved. At the same time, she said, knowing her status could help her find the right person, if he was out there.

"Either way, I was going to get sick," she said. "And I'd want someone who could handle it. If, by some twist of fate, I do get married and have children, at least we know what we're getting into."

After much debate, the friends settled on the third date as the right time to mention Huntington's. But when the first date came, Ms. Moser wished she could just blurt it out.

"It kind of just lingers there," she said. "I really just want to be able to tell people, 'Someday, I'm going to have Huntington's disease.'"

"A Part of My Life"

Last May 6, a year to the day after she had received her test results, the subject line "CAG Count" caught Ms. Moser's attention as she was scrolling through the online discussion forums of the Huntington's Disease Advocacy Center. She knew she had forty-five CAG repeats, but she had never investigated it further.

She clicked on the message.

"My mother's CAG was 43," it read. "She started forgetting the punch line to jokes at 39/40." Another woman, whose husband's CAG count was forty-seven, had just sold his car. "He's 39 years old," she wrote. "It was time for him to quit driving."

Quickly, Ms. Moser scanned a chart that accompanied the messages for her number, forty-five. The median age of onset to which it corresponded was thirty-seven.

Ms. Elio got drunk with her husband the night Ms. Moser finally told her.

"That's twelve years away," Ms. Moser said.

The statistic, they knew, meant that half of those with her CAG number started showing symptoms after age thirty-seven. But it also meant that the other half started showing symptoms earlier.

Ms. Moser, meanwhile, flew to the annual convention of the Huntington's Disease Society, which she had decided at the last minute to attend.

"Mother or father?" one woman, twenty-three, from Chicago, asked a few minutes after meeting Ms. Moser in the elevator of the Milwaukee Hilton. "Have you tested? What's your CAG?"

She was close to getting herself tested, the woman confided. How did it feel to know?

"It's hard to think the other way anymore of not knowing," Ms. Moser replied. "It's become a part of my life."

After years of trying to wring conversation from her family about Huntington's, Ms. Moser suddenly found herself bathing in it. But for the first time in a long time, her mind was on other things. At a youth support group meeting in the hotel hallway, she took her place in the misshapen circle. Later, on the dance floor, the spasms of the symptomatic seemed as natural as the gyrations of the normal.

"I'm not alone in this," Ms. Moser remembers thinking. "This affects other people, too, and we all just have to live our lives."

SEIZING THE DAY

July 15, the day of Ms. Moser's twenty-fifth birthday party, was sunny, with a hint of moisture in the air. At her aunt's house in Long Beach, New York, Ms. Moser wore a dress with pictures of cocktails on it. It was, she and Ms. Elio told anyone who would listen, her "cocktail dress." They drew the quotation marks in the air.

A bowl of "Cure HD" pins sat on the table. Over burgers from the barbecue, Ms. Moser mentioned to family members from her father's side that she had tested positive for the Huntington's gene.

"What's that?" one cousin asked.

"It will affect my ability to walk, talk, and think," Ms. Moser said. "Sometime before I'm fifty."

"That's soon," an uncle said matter-of-factly.

"So do you have to take medication?" her cousin asked.

"There's nothing really to take," Ms. Moser said.

She and the Elios put on bathing suits, loaded the children in a wagon, and walked to the beach.

More than anything now, Ms. Moser said, she is filled with a sense of urgency.

"I have a lot to do," she said. "And I don't have a lot of time."

Over the next months, Ms. Moser took tennis lessons every Sunday morning and went to church in the evening.

When a planned vacation with the Elio family fell through at the last minute, she went anyway, packing Disney World, Universal Studios, Wet 'n Wild and Sea World into thirty-six hours with a high school friend who lives in Orlando. She was honored at a dinner by the New York chapter of the Huntington's society for her outreach efforts and managed a brief thank-you speech despite her discomfort with public speaking.

Having made a New Year's resolution to learn to ride a unicycle, she bought a used one. "My legs are tired, my arms are tired, and I definitely need protection," she reported to Ms. Elio. On Super Bowl Sunday, she waded into the freezing Atlantic Ocean for a Polar Bear swim to raise money for the Make-A-Wish Foundation.

Ms. Elio complained that she hardly got to see her friend. But one recent weekend, they packed up the Elio children and drove to the house the Elios were renovating in eastern Pennsylvania. The kitchen floor needed grouting, and, rejecting the home improvement gospel that calls for a special tool designed for the purpose, Ms. Moser and Ms. Elio had decided to use pastry bags.

As they turned into the driveway, Ms. Moser studied the semi-attached house next door. Maybe she would move in one day, as the Elios had proposed. Then, when she could no longer care for herself, they could put in a door.

First, though, she wanted to travel. She had heard of a job that would place her in different occupational therapy positions across the country every few months and was planning to apply.

"I'm thinking Hawaii first," she said.

Then they donned gloves, mixed grout in a large bucket of water and began the job.

RICHARD PRESTON

An Error in the Code

FROM *THE NEW YORKER*

Just one misspelling among the three billion "letters" that code the human genome can make someone, quite literally, try to tear himself apart. Richard Preston meets two men trying to cope with Lesch-Nyhan syndrome, and the researchers who are trying to understand it.

O NE DAY IN SEPTEMBER 1962, a woman who here will be called Deborah Morlen showed up at the pediatric emergency room of the Johns Hopkins Hospital, in Baltimore, carrying her four-and-a-half-year-old son, Matthew. He was spastic, and couldn't walk or sit up; as an infant, he had been diagnosed as having cerebral palsy and developmental retardation. The emergency room was in the Harriet Lane Home for Invalid Children, an old brick building in the center of the Johns Hopkins complex, where a pediatric resident named Nancy Esterly saw Matthew. He had strange-colored urine, Mrs. Morlen told Esterly, "and there's sand in his diaper." Esterly removed the boy's diaper. It was stained a deep, bright orange, with a pink tinge. She touched the cloth and felt grit. She had no idea what this was, except that the pink looked like blood. She learned from Mrs. Morlen that Matthew had an older brother, Harold, who was also spastic and retarded. Harold was living at the Rosewood State Hospital, an institution for disabled children, outside Baltimore, while Matthew lived at home.

Since both brothers seemed to have the same condition, Esterly thought it likely that they had a genetic disease, but, if so, it wasn't one that she'd ever seen or heard of. Esterly also noted that Matthew was wearing mittens, even though it was a warm day. She admitted the little boy to the hospital.

Esterly took a sample of Matthew's urine, and both she and an intern looked at it under a microscope. They saw that it was filled with crystals. They were beautiful—they were as clear as glass, and they looked like bundles of needles, or like fireworks going off. They were sharp, and it was clear that they were tearing up the boy's urinary tract, causing bleeding. Esterly and the intern pored over photographs of crystals in a medical textbook. The intern asked if the crystals might be uric acid, a waste product excreted by the kidneys; however, cystine, an amino acid that can form kidney stones, seemed the most likely candidate. Esterly needed a confirmation of that diagnosis, so she carried the sample upstairs to the top floor, where William L. Nyhan, a pediatrician and research scientist, had a laboratory. "Bill Nyhan was the guru of metabolism," Esterly told me.

Nyhan, who was then in his thirties, had been studying how cancer cells metabolized amino acids, in an attempt to find ways to cure cancer in children. "It was one of my impossible projects," he said to me recently. Nyhan is now a professor of pediatrics at the University of California, San Diego, School of Medicine. "I love working with kids, but dealing with pediatric cancer was depressing, saddening, and, in truth, maddening," he said. Nyhan ran some tests on Matthew's urine, using equipment he had designed. The crystals weren't cystine, or any sort of amino acid. They proved to be uric acid. A high concentration of uric acid in a person's blood can lead to gout, a painful disease in which crystals grow in the joints and extremities, particularly in the big toe. Gout has been known since the time of Hippocrates, and it occurs mainly in older men. Yet the patient here was a little boy. Nyhan had a medical student named Michael Lesch working in his lab, and together they went downstairs.

Matthew lay in a bed in an open ward on the second floor of the Harriet Lane Home. He was a spot of energy in the ward, a bright-eyed child with a body that seemed out of control. The staff had tied his arms and legs to the bedframe with strips of white cloth, to keep him from thrashing, and they had wrapped his hands in many layers of gauze; they looked like white clubs. Nurses hovered around the boy. "He knew I was a doctor and he knew where he was. He was alert," Nyhan says. Matthew greeted Lesch and Nyhan in a friendly way, but his speech was almost unintelligible: he had dysarthria, an inability to control the muscles that make speech. They noticed scarring and fresh cuts around his mouth.

They inspected Matthew's feet. No sign of gout. Then the boy's arms and legs were freed, and Lesch and Nyhan saw a complex pattern of stiff and involuntary movements, a condition called dystonia. Nyhan had the gauze unwrapped from the boy's hands.

Matthew looked frightened. He asked Nyhan to stop, and then he began crying. When the last layer was removed, they saw that the tips of several of the boy's fingers were missing. Matthew started screaming, and thrust his hands toward his mouth. With a sense of shock, Nyhan realized that the boy had bitten off parts of his fingers. He also seemed to have bitten off parts of his lips.

"The kid really blew my mind," Nyhan said. "The minute I saw him, I knew that this was a syndrome, and that somehow all of these things we were seeing were related."

Lesch and Nyhan began to make regular visits to the ward. Sometimes Matthew would reach out and snatch Nyhan's eyeglasses and throw them across the room. He had a powerful throw, apparently perfectly controlled, and it seemed malicious. "Sorry! I'm sorry!" Matthew would call, as Nyhan went to fetch his glasses.

The doctors persuaded Mrs. Morlen to bring her older son to the hospital. Harold, it turned out, had bitten his fingers even more severely than Matthew, and had chewed off his lower lip. Both boys were terrified of their hands, and screamed for help even as they bit them. The boys' legs would scissor, and they tended to fling out one arm and the opposite leg, like a fencer lunging. The Morlen brothers, the doctors found, had several times more uric acid in their blood than normal children do.

Nyhan and Lesch visited the Morlen home, a row house in a working-class neighborhood in East Baltimore, where Matthew was living with his mother and grandmother. "He was a well-accepted member of his little household, and they were very casual about his condition," Nyhan says. The women had devised a contraption to keep him from biting his hands, a padded broomstick that they placed across his shoulders, and they tied his arms to it like a scarecrow. The family called it the "stringlyjack." Matthew often asked to wear it.

Nyhan and Lesch also discovered that they liked the Morlen brothers. Lesch, who is now the chairman of the Department of Medicine at St. Luke's–Roosevelt Hospital, in New York City, said, "Matthew and Harold were really engaging kids. I enjoyed being around them."

Two years after meeting Matthew Morlen, Nyhan and Lesch published the first paper describing the disease, which came to be called Lesch-Nyhan syndrome. Almost immediately, doctors began sending patients to Nyhan. Very few doctors had ever seen a person with Lesch-Nyhan syndrome, and boys with the disease were, and are, frequently misdiagnosed as having cerebral palsy. (Girls virtually never get it.) Nyhan himself found a number of Lesch-Nyhan boys while visiting state institutions for developmentally disabled people.

When I asked him how long it took him to diagnose a case, he said, "Seconds." He went on, "You walk into a big room, and you're looking at a sea of blank faces. All of a sudden you notice this kid staring at you. He's highly aware of you. He relates readily to strangers. He's usually off in a corner, where he's the pet of the nurses. And you see the injuries around his lips."

WILLIAM NYHAN IS NOW EIGHTY-ONE, a tall, fit man with blondish-gray hair and blue eyes. He has a laboratory overlooking a wild canyon near the UCSD Medical Center. One day when I visited him, two red-tailed hawks were soaring above the canyon, tracing circles in the air. In the years since he identified Lesch-Nyhan, he has discovered or codiscovered a number of other inherited metabolic diseases, and he has developed effective treatments for some of them. He figured out how to essentially cure a rare genetic disorder called multiple carboxylase deficiency, which could kill babies within hours of birth, by administering small doses of biotin, a B vitamin. Lesch-Nyhan, however, has proved to be more intractable.

Decades after the discovery of Lesch-Nyhan syndrome, it is still mysterious. It is perhaps the clearest example of a simple change in the human DNA which leads to a striking, comprehensive change in behavior. In 1971, William Nyhan coined the term "behavioral phenotype" to describe the nature of diseases like Lesch-Nyhan syndrome. A phenotype is an outward trait, or a collection of outward traits, that arises from a gene or genes—for example, brown eyes. Someone who has a behavioral phenotype displays a pattern of characteristic actions that can be linked to the genetic code. Lesch-Nyhan syndrome seems to be a window onto the deepest parts of the human mind, offering glimpses of the mechanics of the genetic code operating on thought and personality.

H. A. Jinnah, a neurologist at Johns Hopkins Hospital, has been studying Lesch-Nyhan syndrome for more than fifteen years. "This is a very horrible disease, and a very complex brain problem," he said. "It is also one of the best models we have for trying to trace the action of one gene on complex human behavior."

A child born with Lesch-Nyhan syndrome seems normal at first, but by the age of three months he has become a so-called floppy baby, and can't hold up his head or sit up. His diapers may have orange sand in them. When the boy cuts his first teeth, he starts using them to bite himself, and he screams in terror and pain during bouts of self-mutilation. "I get calls in the middle of the night from parents, saying, 'My kid's chewing himself to bits—what do I do?'" Nyhan said. The boy ends up in a wheelchair, because he can't learn to walk. As he grows older, his self-injurious behaviors become subtle or more elaborate, more devious. He seems to be possessed by a demon that forever seeks new ways to hurt him. He spits, strikes, and curses at the people he likes the most; one way to tell if a Lesch-Nyhan patient doesn't like you is if he's being nice. ("I got beat up once by Matthew," Lesch told me. He had leaned over the boy and asked him how he was feeling, and Matthew had punched him in the nose.) He eats foods he can't stand; he vomits on himself; he says yes when he means no. This is self-sabotage.

A few hundred boys and men alive in the United States today have been diagnosed as having Lesch-Nyhan syndrome. "I think I know most of them," Nyhan said. One boy, known as J.J., ended up living in Nyhan's research unit for a year, when he was eleven. He was a gregarious child, whose hands seemed to hate him. Over time, his fingers had got inside his mouth and nose and had broken out and removed the bones of his upper palate and parts of his sinuses, leaving a cavern in his face. He had also bitten off several fingers. J.J. died in his late teens; in the past, many Lesch-Nyhan patients died in childhood or their teens, from kidney failure. (Both Morlen brothers died young.) Nowadays, they may live into their thirties and forties, but they are generally frail and often die from infections like pneumonia. Occasionally, a man with the disease flings his head backward with such force that his neck is broken. Many Lesch-Nyhan patients die suddenly and often inexplicably.

A Lesch-Nyhan person may be fine for days, until suddenly his hands jump into his mouth with the suddenness of a cobra strike, and he cries for help. People with Lesch-Nyhan feel pain as acutely as anyone else does, and they are horrified by the idea of their fingers or lips being severed. They feel as if their hands and mouth don't belong to them and are under the control of something else. Some Lesch-Nyhan people have bitten off their tongues, and some have a record of self-enucleation—they have pulled out an eye or stabbed it with a sharp object. When the Lesch-Nyhan demon is dozing, they enjoy being around people, they like being the center of attention, and they make friends easily. "They really are great people, and I think that's part of the disease, too," Nyhan said. Some Lesch-Nyhan people are cognitively impaired, while others are clearly bright, but their intelligence can't be measured easily. "How do you measure someone's intelligence if, when you put a book in front of him, he has an irresistible urge to tear out the pages?" Nyhan asked.

IN 1967, J. EDWIN SEEGMILLER, a scientist at the National Institutes of Health, and two colleagues discovered that in Lesch-Nyhan patients a protein called hypoxanthine-guanine phosphoribosyl transferase, or HPRT, which is present in all normal cells, doesn't seem to work. The job of this enzyme is to help recycle DNA. Cells are constantly breaking down DNA into its four basic building blocks (represented by the letters A, T, C, and G, for adenine, thymine, cytosine, and guanine). This process produces compounds called purines, which can be used to form new code. If HPRT is absent, or doesn't work, then certain purines build up in a person's cells, where they are eventually broken down into uric acid, which saturates the blood and crystallizes in the urine.

In the early 1980s, a group of researchers, led by Douglas J. Jolly and Theodore Friedmann, decoded the sequence of letters in the human gene that contains the instructions for making HPRT. It includes 657 letters that code for the protein. Researchers also began sequencing this gene in people who had Lesch-Nyhan. Each had a mutation in the gene, but, remarkably, nearly everyone had a different one; there was no single mutation that caused Lesch-Nyhan. The mutations had apparently appeared spontaneously in each affected family. And, in the majority of cases, the defect consisted of just one misspelling in the code. For example, an American boy known as D.G. had a single G replaced by an A—one out of the three billion letters of code in the human genome. As a result, he was tearing himself apart.

The HPRT gene is found on the X chromosome. Women have two X chromosomes in each cell, and men have an XY pair. Lesch-Nyhan syndrome is an X-linked recessive disorder. This means that if a bad HPRT gene on one X chromosome is paired with a normal gene on the other X chromosome the disease does not develop. A woman who has the Lesch-Nyhan mutation carries it on only one of her X chromosomes—she doesn't develop the syndrome. Any son that she has, however, will have a 50 percent chance of inheriting the syndrome, and any daughter will have a 50 percent chance of being a carrier. (Examples of this type of disease include hemophilia and a form of red-green color blindness.)

Other genetic mutations have been associated with profound behavioral changes. Rett syndrome, which affects mostly girls, is caused by a mutation in a gene that codes for the MeCP2 protein. People with the syndrome compulsively wring their hands and rub them together as if they were washing them. Children with Williams syndrome have an elfin appearance, an affinity for music and language, and an extreme sensitivity to sound, and are very sociable. Williams syndrome is caused by the deletion of a bit of code from chromosome 7. There is still great uncertainty, however, about how much of a role genes play in major conditions such as depression, bipolar disorder, and borderline personality disorder. Even where there is evidence of a family history of a disease, scientists are unsure how a single gene could choreograph a suite of behaviors. There are roughly twenty-five thousand active genes in the human genome, each with about a thousand to fifteen hundred letters of code. The genome could be thought of as a kind of piano with twenty-five thousand keys. In some cases, a few keys may be out of tune, which can cause the music to sound wrong. In others, if one key goes dead the music turns into a cacophony, or the whole piano self-destructs.

The havoc that the Lesch-Nyhan mutation causes can't easily be undone. Early on, Nyhan tried giving his patients allopurinol, a drug that inhibits the production of uric acid; it is effective with gout. It lowered the uric-acid concentration in Lesch-Nyhan patients, but it didn't reduce their self-injurious actions. The uric acid, it seemed, was another symptom, and not a cause of the behavior. Nyhan has experimented with other treatments, such as soft restraints, which seem to relax patients, and the removal of certain teeth. "I'm profligate with those upper teeth," Nyhan said. Some dentists, though, refuse to extract healthy teeth, even when Lesch-Nyhan syndrome is explained to them.

I told Nyhan that I couldn't imagine what it would be like to live with the disease.

"You could ask someone who has it," he replied.

I FIRST MET JAMES ELROD and Jim Murphy in the winter of 1999. They were living next to each other in rented bungalows in a somewhat marginal neighborhood in Santa Cruz, California. Elrod was then in his early forties, and Murphy was just over thirty. (Murphy died in 2004; Elrod, who is now forty-nine, is one of the oldest living people with Lesch-Nyhan.) The men were clients of Mainstream Support, a private company contracted by the State of California to help people with developmental disabilities live in community settings. Before James Elrod came to Santa Cruz, he lived for eighteen years in a state institution in San Jose. Murphy had lived for most of his life at an institution in Sonoma. Mainstream employees, called direct-care staff, stayed with Elrod and Murphy around the clock, to help them with daily tasks and to make sure they didn't harm themselves. Elrod and Murphy had the authority to hire and fire their assistants and direct their work, though an assistant could refuse an order if he thought that it would put the client in danger.

At that time, Mainstream was run by two men named Andy Pereira and Steve Glenn. "James and Jim are real down-and-gritty guys," Pereira said, the first time I talked to him. "They are not sweet types. They're into fast cars and women." Glenn said that he still had difficulty seeing into the labyrinth of Lesch-Nyhan. "There are these Lesch-Nyhan moments when you feel like you've kind of got it," he said. "James and Jim are pretty good at telling you when they think they're in danger of hurting themselves, but, whenever they're doing something, you always have to ask, Is this James or Jim, or is it Lesch-Nyhan?"

James Elrod has a square, good-looking face, which is marked with scars, and brown, hyperalert eyes. His shoulders and arms are large and powerful, but the rest of his body seems slightly diminished. One day, before he was with Mainstream, an attendant left him alone at dinner for a few minutes. To Elrod's horror, his left hand picked up a fork and used it to stab his nose and gouge it out, permanently mutilating his face. "My left side is my devil side," he told me. When I met him, he wore black leather motorcycle gloves that had been reinforced with Kevlar. If he thought that his left hand was threatening him or someone else, he would grab it or swat it with his right hand. He owns a pickup truck, and his assistants drive him around in it. He used to sell flowers on the Santa Cruz pier, and he carries business cards explaining that he has a rare disease that compels him to hurt himself. "I have injured myself in many ways including my nose, as you can see," the card says. "I will even try to hurt myself by getting into trouble with others." One day, a man bought flowers from Elrod and said, "God bless you." "Eat shit," Elrod replied, and handed the man his business card. While crossing a street in his wheelchair, Elrod has been known to try to roll himself into traffic, yelling, "Slow down, you morons!

Don't you know it's Lesch-Nyhan?" His assistants wrestle him to safety.

Elrod was sitting in front of his house in his wheelchair when I arrived. It was a sunny day. He offered me his right hand to shake. When I gripped his glove, the right index finger collapsed. "You broke my finger!" he gasped. Then he grinned and explained that he didn't have that finger. "Some people get all upset when I do that," he said. "Kids love it. They want to break my finger again."

We chatted for a while. "Hey, Richard—danger," he said.

"What's wrong?"

He cautiously pointed at the pencil I was using to take notes. "Your pencil is scaring me. My hand could grab it and put it in my eye," he said. "You'd better go see my neighbor."

Jim Murphy was sitting in his wheelchair at a table in the living room of his house, and an assistant named Michael Roth cut up pancakes and fed them to him with a spoon. Murphy was a bony man with dark hair and a lean, handsome face. He had a neatly trimmed goatee and a crew cut, and his eyes were mobile and sensitive-looking. His lips were missing. Two of his brothers had also had Lesch-Nyhan, and had died when they were young. "Jimmy will be shy when you first meet him," one of his sisters had told me on the phone. I could expect to hear a lot of swearing, though. "He doesn't mean it," she said. "When he swears at me, I just say, 'I love you, too.'"

That day in Santa Cruz, Murphy stared at me out of the corners of his eyes, with his head involuntarily thrown back and turned away, braced against a headboard. His hands were stuffed into many pairs of white socks, and his chest heaved against a rubber strap that held him in place. He started throwing punches at me, and he kicked at me. He seemed to be enduring his disease like a man riding a wild horse. The wheelchair shook.

I kept back. "It's nice to meet you," I said.

"Fuck you. Nice to meet you." Murphy had a fuzzy but pleasant-sounding voice. His speech was very hard to understand. He looked at Roth. "I'm nervous," he said.

"Do you want to be restrained?" Roth asked.

"Yeah."

Roth placed Murphy's wrists and ankles in soft cuffs fastened with Velcro.

"I'm a little nervous, too," I said, and sat down on the couch.

"I don't care. Good-bye."

I stood up to leave.

Roth explained, however, that this was one of those Lesch-Nyhan situations where words mean their opposite.

Later, Murphy tried to tell me what his disease was like. "You try to tick everybody off, and then you feel bad when you do it," he said. "If you get too close to me, I could—" he said; the ending was indecipherable.

"I'm sorry, what?"

"Coldcock you, Richard. I'll say, 'Get my water,' and I'll give you a sucker punch."

A pair of red boxing gloves hung on the wall. Every day, his assistants placed him on a wrestling mat on the floor, where he rolled around and did stretches and then boxed with them. "I could definitely whip you," he told me. I didn't doubt it.

THERE HAVE BEEN ABOUT twenty autopsies of Lesch-Nyhan patients over the years. Their brains appeared to be perfectly normal. "It's a problem in the connections, in the way the brain functions," H. A. Jinnah, the Johns Hopkins neurologist, said. During some of the autopsies, doctors tested samples of brain tissue to see if they contained normal levels of neurotransmitters—chemicals that are used for signaling between nerve cells. In the Lesch-Nyhan brains, a lemon-size area containing structures called the basal ganglia, near the center of the brain, had 80 percent less dopamine—an important neurotransmitter—than a normal brain. The basal ganglia are wired into circuits that run all over the brain and affect a wide range of functions: motor control, higher-level thinking, and eye movement, as well as impulse control and enthusiasm.

"People with Lesch-Nyhan have an excess number of involuntary movements," Jinnah said. "It's as if they are stepping on the gas too hard when they try to do something. If you ask them to look at a red ball, for instance, their eyes go to everything except the red ball, and they can't explain why. Then, if you introduce a yellow ball into their field of view, but you don't say anything about it, they watch the yellow ball." The moment you draw their attention to it, however, they look away.

"Lesch-Nyhan is at the far end of a spectrum of self-injurious behavior," Jinnah went on. "We all do things that are bad for us. We'll sit down in front of the television and eat a quart of ice cream. We all have self-injurious impulses, too. Driving a car, we can have a strange impulse to drive it the wrong way and smash it into something." Edgar Allan Poe called such promptings "the imp of the perverse." The imp may be signals coming out of the basal ganglia. Normal people feel the promptings of the imp, but most of the time they don't act on them. Lesch-Nyhan may suggest a way in which original thoughts and ideas seem to arise as impulses that aren't suppressed, and how intimate the terrain is between the creative and the self-destructive. "Many people bite their fingernails," Jinnah said. "They'll tell you it's gross and that they don't want to do it—'Sometimes I get nervous and start biting my fingernails,' they'll say. There are people who chew their lips nervously. Now let's turn up the volume a little: some people bite their cuticles. Turn up the volume a little more: some people bite their cuticles until they bleed. Now let's turn the volume *way* up. Now you have someone biting off tissue and bone in his fingers, biting off the whole finger, and chewing his lips off. Where, in this spectrum of behavior, is free will?"

In some ways, Lesch-Nyhan syndrome looks like Parkinson's disease reversed. People with Parkinson's have trouble starting physical actions, and are said to be hypokinetic. Lesch-Nyhan people start actions too easily, and can't stop an action once it starts; they are said to be hyperkinetic. Because Parkinson's is also associated with a deficiency of dopamine in the basal ganglia, scientists have looked to each disease for clues to the other.

In 1973, a researcher named George Breese, at the University of North Carolina School of Medicine, was working with rats that modeled Parkinson's disease. He was treating newborn rats with compounds that changed the dopamine levels in their brains, when, to his surprise, the rats started chewing off their paws. He had inadvertently created a rat with Lesch-Nyhan symptoms. "I'll not go further into the details of what the rats were doing. They weren't biting their mouth tissues, the way human patients do," Breese told me. If he gave the self-injuring rats another compound, they stopped biting their paws—that is, he found a way to reverse the symptoms. "We treated the rat the moment we saw the animal make the first pinprick injury to its paws," he said. The compound, however, has never been approved for use on humans.

IN APRIL 2000, A neurosurgeon at the Tokyo Women's Medical University named Takaomi Taira performed brain surgery on a nineteen-year-old man with Lesch-Nyhan. The young man was living with his parents in a district north of Tokyo. In addition to exhibiting self-injurious behavior, he had the spastic, stiff, thrashing movements of dystonia. "These dystonic movements were getting more severe almost by the day, and his parents were getting desperate," Taira said to me recently. He decided to perform a procedure called deep-brain stimulation to try to calm down the movements.

Deep-brain stimulation was developed by doctors more than twenty years ago for treating people with Parkinson's disease. One or more thin wires are inserted through openings in the skull, and the wires are carefully navigated through the brain until they stop in a part of the basal ganglia called the globus pallidus (the "pale globe"). The wires are connected to a battery pack, which is implanted under the skin of the patient's chest, and a faint, pulsed current of electricity runs through them into the globus pallidus, numbing a spot the size of a pea. The patient feels nothing. The procedure often helps to calm the tremors in Parkinson's patients' hands and limbs, and helps them walk more easily.

"After the surgery, the boy's dystonic movement completely disappeared," Taira said. He sent him home with the deep-brain stimulator, feeling that the operation had helped. Several months later, the young man's parents told Taira that he had stopped biting himself. He was still in a wheelchair, and his uric-acid levels remained high, but he was reading comic books and watching television, and seemed to be enjoying life as never before. "It was completely unexpected, remarkable, almost unbelievable," Taira said. A few years later, the young man suddenly began biting his hands again, and the parents brought him back. "I checked the device and found that the battery was flat. I replaced the battery, and his symptoms were controlled again," Taira said.

A research group in Montpellier, France, led by a neurosurgeon named Philippe Coubes, has given deep-brain-stimulation implants to five Lesch-Nyhan patients. His method involves the insertion of four wires into the brain. "So far, we have three patients who are doing very well and two who are having an intermediate response—the response of one of those is not poor but is not as good as the others," Coubes said. "I'm not sure we will be able to control all their behaviors over the long term, but we are in the process of getting a better understanding of deep-brain stimulation for these patients." The imp of the perverse can be put to sleep, but nobody knows how to make it go away.

Scientists aren't sure why deep-brain stimulation seems to work in some patients, or if it can help others; indeed, the results are a reminder of how obscure the workings of the brain still are. Nor is it clear what the risks might be. William Nyhan was cautious about the procedure's potential. "I see these kids as fragile, and they don't respond very well to surgical invasions," he said.

At Johns Hopkins, though, Jinnah was anxious to begin a study on a group of at least eight Lesch-Nyhan patients using deep-brain stimulation. He still needs to secure funding and get approval from the federal government. (The procedure has not been specifically approved for Lesch-Nyhan patients.)

Jinnah has never had an easy time getting funding and attention for Lesch-Nyhan research. He says, "People ask me, 'Why not study more common diseases?' My answer is that if we neurologists did that we'd all be studying Alzheimer's disease, Parkinson's disease, and strokes. There are thousands of other brain diseases out there, and they're all orphans. But these rare diseases may teach us something new about the brain, something relevant to the common brain diseases which affect so many people."

I WENT BACK SEVERAL TIMES to visit James Elrod and Jim Murphy, and began helping their staff with daily tasks. Elrod spat in my face a few times, and gave me a left jab to the jaw. Once, his Kevlar-covered fingers closed on my skin like pliers; he apologized while we both worked to get them loose. Murphy, at his thirty-third-birthday party, planted his face in his cake, and then punched me. Nevertheless, I came to like them a lot. Murphy had a passion for off-road driving, which he was not usually able to indulge. One day, in 2001, I showed up in Santa Cruz in a rented Ford Expedition with four-wheel drive. An assistant named Tracye Overby was with Murphy, while another, named Chris Reeves, was assigned to Elrod. I drove the group to a dried-out lake bed near Watsonville called College Lake, which we'd heard was a good place for four-wheeling. On the way, I stopped to get directions from a California state trooper. "I would not advise going there with disabled people," he said.

College Lake turned out to be a mile-wide expanse of lardlike clay, covered with sand. The lake bed had a dark, wet-looking center, and was surrounded by thickets of willows. I edged the Expedition out onto the sand. "Go faster," Murphy said. "You're driving like an old lady." I gunned the engine, the Expedition leaped forward, and we raced across the sand. When I turned sharply, the men roared with delight. I performed a figure eight, then aimed the vehicle toward the center of the lake and ran it up to full power. We passed the sunken carcass of a truck, buried up to its roof in clay. The Expedition slowed. Then it began tipping over, and I realized that we were driving across what you might call quickmud. If we stopped, we would go down. I floored the engine, but it was too late. The wheels began spinning, we came to a halt and sank up to the doors, and the engine stalled.

There was a moment of silence, and then Elrod and Murphy erupted with obscenities directed at me. The two helpers seemed unperturbed. "This is just the nature of our work," Reeves said. "Everything that you plan never goes as you planned it."

After several tries on my cell phone, I reached a towing company that was willing to try to get us out, but it would need to be paid in advance, cash preferred, no results guaranteed.

"I'm nervous," Murphy said as we waited. Blood dribbled out of his mouth—he was biting himself. Overby lifted him out of the car, carried him across the sand to the shade of some willows, and sat down, holding him in her lap. She wiped his mouth with a napkin, cradled his head in her arms, and began singing to him. He began to laugh.

Elrod, sitting in the front seat of the Expedition, began laughing, too. The men were connoisseurs of what I had done: I had ignored the advice of a police officer and driven two disabled men at high speed into the mud. They saw something familiar in my behavior.

Three years later, in 2004, Murphy came down with pneumonia. When it became clear that he was dying, I called him to say good-bye. As he came on the line, I could hear voices in the background; more than thirty people had come to see him. "I'll be all right," he said, and added, "Take care with your driving."

Another day, before Murphy died, I visited James Elrod. Tracye

Overby, who was working as Elrod's assistant, needed to change the silk liners that he wears inside his motorcycle gloves.

Elrod did not like to see his bare hands. He asked me to hold his wrists while Overby removed his gloves. The hands that emerged were pale, with spindly fingers that had been gnawed close to the bone in places, and a finger was missing. "Danger," he said. His eyes took on a strange, bright, blank look. He was staring at the right hand. His arm was tense and trembling. As if a magnet were pulling it, the hand moved toward his mouth. "Help!" he called in a muffled voice.

We threw ourselves on Elrod. It took all our strength to restrain his hand. As soon as we got control of it, he relaxed. Overby got the gloves back on.

"Nobody knows about this disease. Every day I'm hoping for a cure," Elrod said. "I wanted you to see that."

Thomas Goetz

23andMe Will Decode Your DNA for $1,000. Welcome to the Age of Genomics

FROM *WIRED*

It's a Silicon Valley start-up like no other: a company that allows you, for a fee, to investigate your DNA for clues about your ancestry and risk for disease. Thomas Goetz catches up with the two women who founded the company—and who may be transforming the health industry.

A T THE AGE OF SIXTY-FIVE, my grandfather, the manager of a leather tannery in Fond du Lac, Wisconsin, suffered a severe heart attack. He had chest pains and was rushed to the hospital. But that was in 1945, before open heart surgery, and he died a few hours later. By the time my father reached sixty-five, he was watching his diet and exercising regularly. That regimen seemed fine until a couple of years later, when he developed chest pains during exercise, a symptom of severe arteriolosclerosis. A checkup revealed that his blood vessels were clogged with arterial plaque. Within two days he had a triple bypass. Fifteen years later (fifteen years that he considers a gift), he's had no heart trouble to speak of.

I won't reach sixty-five till 2033, but I have long assumed that, as regards heart disease, my time will come. My genes have predetermined it. To avoid my father's surgery, or my grandfather's fate, I try to eat healthier than most, exercise more than most, and never even consider smoking. This, I figure, is what it will take for me to live past sixty-five.

Turns out that my odds are better than I thought. My DNA isn't pushing me toward heart disease—it's pulling me away. There are established genetic variations that researchers associate with a higher risk for a heart attack, and my genome doesn't have any of those negative mutations; it has positive mutations that actually reduce my risk. Like any American, I still have a good chance of eventually developing heart disease. But when it comes to an inherited risk, I take after my mother, not my father.

Reading your genomic profile—learning your predispositions for various diseases, odd traits, and a talent or two—is something like going to a phantasmagorical family reunion. First you're introduced to the grandfather who died twenty-three years before you were born, then you move along for a chat with your parents, who are uncharacteristically willing to talk about their health—Dad's prostate, Mom's digestive tract. Next, you have the odd experience of getting acquainted with future versions of yourself, ten, twenty, and thirty years down the road. Finally, you face the prospect of telling your children—in my case, my eight-month-old son—that he, like me, may face an increased genetic risk for glaucoma.

The experience is simultaneously unsettling, illuminating, and empowering. And now it's something anyone can have for about $1,000. This winter marks the birth of a new industry: companies will take a sample of your DNA, scan it, and tell you about your genetic future, as well as your ancestral past. A much-anticipated Silicon Valley start-up called 23andMe offers a thorough tour of your genealogy, tracing your DNA back through the eons. Sign up members of your family and you can track generations of inheritance for traits like athletic endurance or bitter-taste blindness. The company will also tell you which diseases and conditions are associated with your genes—from colorectal cancer to lactose intolerance—giving you the ability to take preventive action. A second company, called Navigenics, focuses on matching your genes to current medical research, calculating your genetic risk for a range of diseases.

The advent of retail genomics will make a once-rare experience commonplace. Simply by spitting into a vial, customers of these companies will become early adopters of personalized medicine. We will not live according to what has happened to us (that knee injury from high school or that twenty pounds we've gained since college) nor according to what happens to most Americans (the one-in-three chance men have of getting cancer, or women have of dying from heart disease, or anyone has for obesity). We will live according to what our own specific genetic risks predispose us toward.

This new industry draws on science that is just beginning to emerge. Genomics is in its earliest days: The Human Genome Project, the landmark effort to sequence the DNA of our species, was completed in 2003, and the research built on that milestone is only now being published. The fact that any consumer with $1,000 can now capitalize on this project is a rare case of groundbreaking science overlapping with an eager marketplace. For the moment, 23andMe and Navigenics offer genotyping: the strategic scanning of your DNA for several hundred thousand of the telltale variations that make one human different from the next. But in a few years, as the price of sequencing the entire genome drops below $1,000, all six billion points of your genetic code will be opened to scrutiny.

To act on this data, we first need to understand it. That means the companies must translate the demanding argot of genetics—alleles and phenotypes and centromeres—into something approachable, even simple, for physicians and laypersons alike. It's one thing for a doctor to tell patients that smoking is bad for them, or that their cholesterol count is high. But how are you supposed to react when you're told you have a genetic variation at rs6983267 that's been associated with a 20 percent higher risk of colorectal cancer? And what are physicians, most likely untrained in and unprepared for genomic medicine, to do when a patient comes in wielding a printout that indicates a particular variation of a particular gene?

This new age of genomics comes with great opportunity—but also great quandaries. In the genomic age, we will no longer have the problem of not knowing, but we will face the burden of whether we want to know in the first place. We'll learn what might be best for us in life and then have to reckon with the risk and perhaps the guilt of not acting on that knowledge. We will, counterintuitively, face even more pressure to conduct our lives carefully, strictly, and cautiously; we'll practice the art of predictive diagnosis and receive a demanding roster of things to avoid, things to do, and treatments to receive—long before there's any physical evidence of disease. And, yes, we will know whether our children are predisposed to certain traits or talents—athletics or music or languages—and encourage them to pursue certain paths. In short, life will become a little more like a game of strategy, where we're always playing the percentages, trying to optimize our outcomes. "These are enormously large calculations," says Leroy Hood, a pioneer of genomic sequencing and cofounder of the Institute for Systems Biology in Seattle, who suggests that if we pay attention and get the math right, "it's not a stretch to say that we could increase our productive lifespans by at least a decade."

THE QUESTION WAS SURELY STRANGE. In February 2005, Anne Wojcicki sat down at the so-called Billionaires' Dinner, an annual event held in Monterey, California, and asked her tablemates about their urine. She was curious whether, after eating asparagus, they could smell it when they urinated. Among those at her table were geneticist Craig Venter; Ryan Phelan, the CEO of DNA Direct, a San Francisco genetic-testing company; and Wojcicki's then-boyfriend (and now husband), Sergey Brin, cofounder of Google. Most could pick up the smell of methyl mercaptan, a sulfur compound released as our guts digest the vegetable. But some had no idea what Wojcicki was talking about. They had, it seems, a genetic variation that made the particular smell imperceptible to them.

Soon, the conversation turned to a growing problem: While researchers are amassing great knowledge about certain genes and genetic variations, there is no way for people to access that data for insights about themselves and their families—to Google their genome, as it were. As a biotech and health care analyst at Passport Capital, a San Francisco hedge fund firm, Wojcicki knew that the pharmaceutical industry was already at work on tailoring drugs to specific genetic profiles. But she was intrigued by the prospect of a database that would compile the available research into a single resource.

Linda Avey wasn't at the dinner, but she wished she had been when she read about it later that year in David Vise and Mark Malseed's book, *The Google Story.* At the time, Avey was an executive at Affymetrix, the company that had pioneered some of the tools for modern genetic research. For nearly a year, she had been mulling the idea of a genotyping tool for consumers, one that would let them plumb their own genome as well as create a novel data pool for researchers. She even had a placeholder name for it: Newco. "All the pieces were there," Avey says. "All we needed was the money, as usual, and computational power." Two things that Google has plenty of. Around the time she read Vise and Malseed's book, Avey had a dinner scheduled with a Google executive. She asked Wojcicki to join them, and the two quickly hit it off. Within a few months, they had settled on the idea behind 23andMe: Give people a look at their genome and help them make sense of it. (The company's name is a reference to the twenty-three pairs of chromosomes that contain our DNA.)

Brin offered to be an angel investor. "Sergey was like, 'Come up with something in three months and launch it,'" Wojcicki says. "We thought it would be so fast." In fact, the project took more than eighteen months from conception to launch. Last spring, Google invested $3.9 million in 23andMe (part of the proceeds repaid Brin, who has since recused himself from the investment). The company, which now has more than thirty employees in a building down the road from Google, feels very much like the quintessential start-up. In the entry hall, alongside two Segways (a gift from inventor Dean Kamen), stands a herd of pedal-pusher bicycles. On a whiteboard in the hall, someone has scrawled an anxiety meter. *Current threat level: slight deformation* (engineering-speak for moderate stress). But that level had been crossed out and the alert upped to *bananas.*

Still, 23andMe is hardly a typical Valley outfit. Instead of widgets and Ajax apps, the cubicle chatter more likely concerns Klinefelter's syndrome and hermaphrodites. Such banter underscores a major challenge for the company: making customers comfortable with the strange vocabulary and discomfiting implications of genetics. As Avey notes, when you're asking your customers for their spit, best to have an especially strong relationship.

A lot of spit, as it turns out. It takes about ten minutes of slavering to fill the 2.5-milliliter vial that comes in the fancy lime box provided by 23andMe. Wrap it up, call FedEx, and two to four weeks later you get an e-mail inviting you to log in and review your results. There are three main sections to the Web site: Genome Labs, where users can navigate through the raw catalog of their twenty-three pairs of chromosomes; Gene Journals, where the company correlates your genome with current research on a dozen or so diseases and conditions, from type 2 diabetes to Crohn's disease; and Ancestry, where customers can reach back through their DNA and discover their lineage, as well as explore their relationships with ethnic groups around the world. Family members can share profiles, trace the origin of particular traits, and compare one cousin's genome to another in a fascinating display of DNA networking. Avey herself has had roughly thirty members of her extended family genotyped, spanning four generations. The effort has turned her clan into what is likely the most thoroughly documented gene pool in the world.

It's the Gene Journals, though, that could really change people's lives. Here customers learn their personalized risk for a particular condition, calculated according to whether their genotype contains markers that research has associated with specific risks. Wojcicki stresses, though, that 23andMe's results are not a diagnosis. "It's simply your information," she insists. In part, this distinction is to make sure the company doesn't run afoul of the Food and Drug Administration, which strictly regulates diagnostic testing for disease but has been slow to respond to the more transformational aspects of genomics. But the caveat also matters because the influence of genetics varies from disease to disease; some conditions have a strong heritable component, while others are determined more by environmental factors.

With its emphasis on disease risks, Navigenics is more comfortable offering something closer to a diagnosis. "If I tell you you've got a genetic likelihood of getting colon cancer, you're going to get a colonoscopy early," says Navigenics cofounder David Agus, a prominent oncologist and director of the Spielberg Family Center for Applied Proteomics at Cedars-Sinai Medical Center in Los Angeles. "And that's going to save lives."

Both companies draw a good lesson from the bad example of the body scan industry. When storefront CT scanning machines popped up in the late 1990s, the idea seemed golden to many radiologists and entrepreneurs: Customers could go directly to an imaging center and get an early look at possible tumors or polyps for about $1,000. But the market cratered by 2005, when it became evident that insurers wouldn't pay for the scan without a prior diagnosis and customers wouldn't pay out of pocket for frequent scans. What's more, the false-positive rate was jarringly high, and anxious customers often raced back to their doctor with an image showing, for instance, benign kidney or liver cysts, only to be told that they were harmless incidental lumps.

In other words, there was too much noise and not enough signal. So both 23andMe and Navigenics are determined not to simply shovel along raw research, with scary one-off results indistinguishable from well-established correlations. Inhouse experts at both companies have filtered and vetted hundreds of studies; only a handful are deemed strong enough to incorporate into their library of conditions, which is used for personalized risk calculations. The hope is that this will reduce or eliminate false alarms and let customers trust the experience—maybe even enjoy it.

One afternoon I was working up my own 2.5 milliliters of spit at the company's office when Jimmy Buffett dropped by to get an early peek at his results. A few month's earlier, the singer had let 23andMe peruse his genotype and compare his genealogy to Warren Buffett's. The two men had long wondered if they were somehow related (they aren't, it turns out). Now Jimmy wanted to check out the whole experience. He sat down in front of a laptop in Wojcicki's office, and she looked over his shoulder, guiding him through the site. First he clicked through his ancestral genome, noting that his maternal lineage showed a strong connection to the British Isles. "So the women came over with the Saxon invasion; pretty cool," he said. Another click and he perused his similarity to other ethnic groups, spotting a strong link to the Basque region of Spain. "No wonder I like Basque food so much," he noted.

Then he clicked over to see his disease risks—and was transfixed. "Wow. Right, that's about right for my family," he said as he ran through various conditions. After about forty-five minutes of self-discovery, he leaned back in his chair to put it all together. "Boy, this can get pretty fascinating. And every time some research comes out, I can log on and see how it works for me. I get it," Buffett said with a laugh. "You guys are mad scientists."

Gregor Mendel began growing peas in his abbey garden in the 1850s, just a simple monk curious about the differences among the plants. A member of the Augustinian order, Mendel took to his garden experiments with characteristic discipline and rigor. He grew some peas with green seeds and others with yellow ones, some with violet flowers and others with white, some with round seed pods and others with wrinkled pods, and so on—at least ten thousand plants in all. By the time he was done, he had established the principles of genetic inheritance, identifying some traits as dominant and others as recessive. (Less celebrated is his later work breeding honeybees; though his hybridized African and South American bees produced wonderful honey, they were exceptionally vicious, and he destroyed them.)

More than a century later, Mendel's basic concepts remain the cornerstone of genetics. We now understand his traits as genes, and genes as sections of DNA—a strand of three billion pairs of ATGC (adenine and thymine, and guanine and cytosine), the nucleotides that compose our genome.

Since 1983, when the gene associated with Huntington's disease was first linked to a particular chromosome, most genetic discoveries have worked like Mendel's peas: They have focused on traits associated with single genes. These so-called monogenic conditions— diseases like hemochromatosis (where the body absorbs too much iron) or Huntington's disease—are easy to research, because the associations are pretty much binary. If you have the genetic mutation, you're almost certain to develop the disease. That makes them easy to screen for, too. There are now tests for more than fourteen hundred of these diseases: prenatal screening for cystic fibrosis, mutations in BRCA1 and BRCA2 genes that convey a strong risk for breast cancer, and so forth. This is the sort of genetic testing most of us are familiar with. And such screening can be extremely useful. Careful testing for Tay-Sachs disease among Ashkenazi Jews, for instance, has led to a 90 percent reduction in the disease in the U.S. and Canada.

But as genetic research has progressed, the idea that most diseases will have a clearly defined, single genetic component—what's known as the "common disease, common gene" hypothesis—has turned out to be mostly wishful thinking. In fact, the fourteen hundred conditions that are currently tested for represent about 5 percent of diseases in developed countries, meaning that for 95 percent of diseases there's something more complicated going on.

Most conditions, it turns out, develop from a subtle interplay among several genes. They are said to be multigenic, not monogenic. And while scientists have made progress connecting the deterministic dots between rare genes and rare conditions, they face a far greater challenge understanding the subtler genetic factors for those more common conditions that have the major impact on society. "We're learning plenty about the molecular basis of disease—that's the revolution right now," says Eric Lander, founding director of the Broad Institute and one of the leaders of the Human Genome Project. "But whether that knowledge translates into personalized predictions and personalized therapeutics is unknown." In other words, not all genes are as simple to understand as Mendel's peas.

The source of this complexity lies in our SNPs, or single nucleotide polymorphisms, the single-letter mutations among the base pairs of DNA—swapping an A for a G, or a T for a C—that largely determine how one human is genetically different from another. Throughout our six billion bits of genetic code, there are millions of SNPs (pronounced "snips"), and some untold number of those play a role in our predilection for disease. For researchers like Lander, the main challenge is establishing which SNPs—or which constellation of SNPs—affect which conditions.

Consider, for instance, the many ways that a human heart can go bad. The arteries supplying blood to the heart can be clogged with plaque, constricting blood flow until the organ goes into arrest. Or a valve in the heart can leak, spilling blood into the lungs and causing pulmonary edema. Or the tissues of the organ itself can be weakened, as in cardiomyopathy, so that the muscle fails to pump enough blood throughout the body. Each of these conditions has specific terminology, causes, and treatments, but they are all versions of heart disease, which is the leading killer in the U.S. And each condition may have its own genetic component, or be influenced by a range of genetic components, with each case of the illness a unique combination of genetic variables and environmental factors. So establishing the genetic component of heart disease means, in actuality, accounting for a daunting variety of conditions and tracking the influence of a broad number of genetic variations, as well as separating them from environmental components.

Now, thanks to a series of complementary innovations, geneticists have begun teasing apart the complexity. First, the Human Genome Project, completed in 2003, provided a map for our common genomic sequence. Next, 2005 saw the completion of the first phase of the International HapMap Project, a less-celebrated but equally ambitious effort that cataloged common patterns of genetic variations, or haplotypes, SNP by SNP. That helped researchers know where they should focus their attention. And finally, by mid-2006 the price of genotyping microarrays—the matchbox-sized chips that can detect SNP variations from genome to genome—had dropped to a level that let scientists greatly increase the pace and scope of their research.

As these three factors have converged, the pace of discovery has taken off, producing a startling number of new associations between SNPs and disease. Even the sober *New England Journal of Medicine* described trying to keep up with the research as "drinking from the fire hose." Lander calls it a twenty-year dream coming to fruition. "2007 has been one of those magical years where the entire picture comes into focus. Suddenly we have the tools to apply to any problem: cancer, diabetes—a huge list of diseases. It's just a stunning explosion of data. Pick a metaphor: we've now landed on this new continent, and the people are out there exploring it, and we're finding mountains and waterfalls and rivers. We're turning on lights in dark rooms. We're finding pieces to the jigsaw puzzle."

Clearly, this is an exciting time to be a geneticist. And, it turns out, a consumer, too.

COME LATE SEPTEMBER, Avey and Wojcicki invited their board of scientific advisers to Mountain View, California, for one last review of the site before launch. The meeting began around noon. Avey, as is her habit, had been going strong since four o'clock that morning. Wojcicki was less sprightly, having just returned the previous night from her three-week honeymoon with Brin on safari in Africa and sailing around Greece and Turkey; she was also coming down with a nasty cold. After some idle chat about the biology of sleep, the board watched a demonstration of the company's user interface. Soon, the discussion turned to the thorny question of how much 23andMe will have to teach its customers about genetics to enable them to understand its offerings. "If we can get them to understand LD, that'll be an accomplishment," Avey said, referring to "linkage disequilibrium," a fairly obscure term describing how some genetic variations occur more often than anticipated. No, said Daphne Koller, a Stanford computer scientist and 2004 MacArthur fellow. "This should be a black box. LD is just going to trip them up."

As it happens, because 23andMe is a Web-based company, it can do both, letting the genetics hobbyist geek out on the details while giving the novice a minimum of information. Still, the challenge here was palpable: starting a personal genomics company isn't like starting a Flickr or a Facebook. There's nothing intuitive about navigating your genome; it requires not just a new vocabulary but also a new conception of personhood. Scrape below the skin and we're flesh and bone; scrape below that and we're code. There's a massive amount of information to comprehend and fears to allay before customers will feel comfortable with the day-to-day utility of the site. 23andMe's solution is to offer a deep menu of FAQs, along with some nifty animation that explains the basic principles of genetics.

But the start-up is also careful not to overwhelm customers with foreboding information. Take its approach to monogenic conditions like Huntington's disease. For one thing, the company makes it clear that it is not in the diagnostic business and therefore doesn't provide specific genetic tests for specific diseases. But even if 23andMe wanted to, the SNP technology doesn't allow it, since many of the fourteen hundred monogenic conditions are diagnosed using techniques other than SNP testing. The BRCA1 and BRCA2 mutations that carry a high risk for breast cancer, for instance, are not SNPs but more complex defects that show up only in a test that sequences the entire gene. Similarly, the test for Huntington's looks for repeats of a certain nucleotide sequence, rather than single-letter variations. Given the rarity of such conditions, it would be cost-prohibitive to include these tests in a $1,000 run.

In other circumstances, the science is evolving so fast that 23andMe must invent a methodology as it goes. Take the essential task of calculating a customer's genetic risk for a disease, which the company delivers under its Odds Calculator. For a condition like type 2 diabetes, at least eight different SNPs have been correlated to the disease. Research among people of European descent has found that each of those SNPs has a slightly different effect—a variation of rs4712523 can increase one's risk by 17 percent, while a variation at rs7903146 can decrease risk by 15 percent. To crunch these numbers and determine one person's risk factor, 23andMe has opted to multiply the risks together. But a competing school of thought argues for adding the risk from SNP to SNP. The two approaches can result in wildly different tallies. "A lot of this is unknown. It's totally experimental," Wojcicki told me a few weeks before the science board meeting. "No one has looked at all eight diabetes markers together. They've all been identified individually, but they don't know exactly how they work together. So we've tried to make that clear."

All the ambiguity is indeed clear. There's no lack of caveats and in-context explanations on the site counseling customers to be cautious. In fact, the board at times even urged the company to hedge less and embrace the technology's gee-whiz factor, including uncertainty, more decisively. George Church, the Harvard geneticist who pioneered the sequencing techniques behind the Human Genome Project, sketched out a scenario: when a new study reporting a genetic association with a disease shows up in the *New York Times*, people are going to log on to 23andMe that morning and check to see whether the genetic marker in question is in their results. "People are going to wonder if you've got them covered," Church said. "And the answer better be yes."

In fact, that answer depends on the DNA chip that 23andMe uses to scan customer genomes. The company outsources that work to Illumina, the chip's developer. In its lab, Illumina extracts DNA from saliva and disperses it across a three-by-one-inch silicon wafer studded with more than 550,000 nanoscopic protein dots. Each dot detects a different SNP; more than half a million dots, strategically distributed across the human genome, cover a meaningful swath of anybody's DNA.

But it's possible that new research could turn up an association with a SNP that the 23andMe scan doesn't look for. And by definition, genotyping is a strategic, rather than an exhaustive, catalog.

The real endgame, therefore, is whole-genome sequencing, where you don't have to hope that you're covered—you'll know it. With whole-genome sequencing, all three billion base pairs of DNA will be identified: a complete library of your genetic code. As with DNA chips, sequencing technology is getting faster and costs are dropping. The Human Genome Project spent nearly three billion dollars to sequence the first human genome. Sequencing DNA codiscoverer James Watson's genome cost just under one million dollars; Craig Venter, who has already sequenced his genome at least once, is now spending about $300,000 to have it read again. Prices are expected to fall even more rapidly now that the X Prize Foundation has offered a ten-million-dollar award to the first team to sequence one hundred human genomes in ten days for less than $10,000 each.

At the board meeting, as talk turned to whole-genome sequencing, the energy in the room picked up. "This is absolutely the future," said Michael Eisen, a computational biologist at the University of California, Berkeley. "It's exactly what the company should be doing as soon as possible."

"We will," responded Wojcicki, who then offered a juicy detail to the board. "We already have ten people lined up and willing to pay $250,000 each for their whole genome. It's definitely something we want to do, maybe even in '08."

"George, how much will $250,000 get you?" Eisen asked Church, who's also on the X Prize advisory board. "How good a sequence would that be?"

"As good as Watson's," Church said. "At least as good."

Pushing the science forward is also a key part of the 23andMe business plan. As the company builds up its roster of customer genotypes, and later whole sequences, it gains a treasure trove of data that in turn can drive further research. On signing up, customers agree that their data, though still confidential, may be made available for scientific purposes. As the pool of participants grows, the startup hopes to forge partnerships with academics and advocacy groups that focus on specific conditions. Already, the Parkinson's Institute is working with 23andMe on a study of Parkinson's disease. Similarly, 23andMe is talking with Autism Speaks, an advocacy group, about initiating research into autism—a disorder so complex that it will require the genetic information of many thousand research subjects to tease out potential associations.

This is also where a novel use of social-networking tools comes in. Wojcicki envisions groups of customers coming together around shared genotypes and SNPs, comparing notes about their conditions or backgrounds and identifying areas for further scientific research on their own. "It's a great way for individuals to be involved in the research world," Wojcicki says. "You'll have a profile, and something almost like a ribbon marking participation in these different research papers. It'll be like, 'How many *Nature* articles have you been part of?'" (Social networking will be included in version 2.0 in a matter of months, Avey says.)

For the board, such enterprising approaches to research are part of the fun of 23andMe. But after a long afternoon in a stuffy conference room, even geneticists can tire of too much genetics, and the meeting wound down. As the group walked into the foyer, someone asked about the two Segways there. Soon enough, some of the world's most celebrated geneticists had hopped aboard and were taking turns racing around the office at top speed.

MY RISK FOR HEART DISEASE may be lower than average, but that doesn't mean my genome isn't primed for problems. Far from it. Variations of three SNPs double my risk for prostate cancer, leaving me with a 30 percent chance of developing it in my lifetime. Restless legs syndrome, a dubious-sounding ailment characterized by jerky twitches in the middle of the night, was recently associated with a particular SNP variation—and I've got it, raising my risk by 32 percent. And my risk for exfoliating glaucoma, a type of eye disease, is a whopping three times the average American's. While the average person has just a 4 percent risk, my risk factor of 12 percent means it's something to mind.

Scanning my spreadsheet, all the odds start looking more like land mines. An 18 percent risk festers for this potentially fatal condition, a 13 percent risk ticks for that debilitating condition, and somewhere out there looms a 43 percent chance for something I may survive but sure don't want. And suddenly I realize: I can try to improve my odds here and there—eat less steak, schedule that colonoscopy earlier than most—but I'm going to go somehow, sometime. I can game the numbers, but I can't deny them.

Think of it this way: health is an equation, with certain inputs and outputs. With conventional medicine, that means some fairly basic algebra: the simple addition and subtraction of symptoms and causes, with treatments like pharmaceuticals and surgery on the other end of the formula. For most Americans, the calculation results in fairly good health, with a lifespan stretching into the seventies. With the advent of genomics, though, we have stumbled into a far more arduous calculus, one requiring a full arsenal of algorithms and vectors. It's a more powerful tool—but it's also a lot more complicated.

It's not just the matter of accounting for all of our genetic markers and computing the attendant risk. That's just the start of it. Real personalized medicine must take into account traditional environmental factors, like smoking and diet and exercise. It also must consider the legion of pathogens out there, each with its own genetic quirks—not only the conventional ones of infectious disease but also the emerging class of viruses that seem to influence conditions from certain cancers to ulcers to obesity. Then there is the microbiome, the trillion-cell ecosystem of microbes that lives inside all of us, contributing to our health in largely mysterious ways. Oh, and save a piece of the equation for epigenetics, changes to the ways genes function without changes in the actual gene sequence. They contribute to our risk for common diseases such as cancer, heart disease, and diabetes.

Finally, leave a big blank spot for chance. No matter how much we learn from our genome, no matter how much it explains about us, randomness is always a looming factor in any health equation. Consider one behavior that is strongly associated with bad health—smoking. Everyone knows smoking is the single worst choice most people can make for their health. Yet the truth is that about a quarter of long-term smokers will not die of a smoking-related disease. Fate doesn't always work in our favor, though: Account for every known risk factor for heart disease—from high cholesterol to smoking to high blood pressure—and that explains only half the cases of the disease in the U.S. In other words, I can bank on my genes and live in the most optimal way . . . and still die of a heart attack.

Mathematics isn't just a metaphor here. All of these variables are being broken down into data by scientists, and each data set is being scrutinized in an effort to quantify its impact on health. So let's make the leap of faith. The science is there, the data has been crunched, and it's all clear: Your genome is telling you that you face an elevated risk for certain diseases. What do you do? First, you likely go to your doctor (and let's assume she is one of the mere eight hundred M.D.s nationwide who has some training in genetics, so that she can actually make sense of your information). She considers your elevated risk and recommends some specific changes to your lifestyle. Will that work?

It might, if you act on that advice. But odds are you won't. In 1981, the National Institutes of Health completed a ten-year study that stands as the largest effort in scientific history to track behavior change. Starting with a pool of more than 360,000 Americans, the NIH set up centers around the country to study how well people would follow behaviors to alleviate the risk of heart disease. The subjects received personal counseling and support to help them stop smoking, eat better, and lose weight. At the study's end, though, 65 percent of the smokers still had the habit, half of those with high blood pressure still had it, and few had changed their diet at all. Subsequent studies have shown the same thing: Changing behavior is hard.

Luckily, there will be drugs tailored to work more effectively with our genetic quirks. These pharmacogenomics already exist: Herceptin specifically targets breast cancers that are caused by a growth protein from the HER2 gene, for instance, and more are in development. But taking a drug for several years, even one tailored to your DNA, can create a new set of disease risks and initiates a new trajectory of calculations.

The question becomes, then, whether you want to embark on this path of oddsmaking in the first place. Many individuals won't want to know what their genome has in store. Others will, only to join the worried well—those who live in fear of fulfilling their genetic destiny. And, of course, those genotyped or sequenced at birth won't have that choice; it'll already have been made for them.

Still, Wojcicki is onto something when she describes our genome as simply information. Already, we calibrate our health status in any number of ways, every day. We go to the drugstore and buy an HIV test or a pregnancy test. We take our blood pressure, track our cholesterol, count our calories. Our genome is now just one more metric at our disposal. It is one more factor revealed, an instrument suddenly within reach that can help us examine, and perhaps improve, our lives.

CARL ZIMMER

Evolved for Cancer?

FROM *SCIENTIFIC AMERICAN*

Human evolution may have produced some natural defenses against cancer. Carl Zimmer explores why these defenses may lose their effectiveness as we age—and finds evolution at work there as well.

NATURAL SELECTION IS NOT NATURAL PERFECTION. Living creatures have evolved some remarkably complex adaptations, but we are still very vulnerable to disease. Among the most tragic of those ills—and perhaps most enigmatic—is cancer. A cancerous tumor is exquisitely well adapted for survival in its own grotesque way. Its cells continue to divide long after ordinary cells would stop. They destroy surrounding tissues to make room for themselves, and they trick the body into supplying them with energy to grow even larger. But the tumors that afflict us are not foreign parasites that have acquired sophisticated strategies for attacking our bodies. They are made of our own cells, turned against us.

Nor is cancer some bizarre rarity: a woman in the U.S. has a 39 percent chance of being diagnosed with some type of cancer in her lifetime. A man has a 45 percent chance.

These facts make cancer a grim yet fascinating puzzle for evolutionary biologists. If natural selection is powerful enough to produce complex adaptations, from the eye to the immune system, why has it been unable to wipe out cancer? The answer, these investigators argue, lies in the evolutionary process itself. Natural selection has favored certain defenses against cancer but cannot eliminate it altogether. Ironically, natural selection may even inadvertently provide some of the tools that cancer cells can use to grow.

The study of cancer evolution is still in its infancy, with much debate about the mechanisms involved and much testing of hypotheses left to carry out. Some medical researchers remain skeptical that the work will affect the way they fight the disease. Evolutionary biologists agree that they are not about to discover a cure for cancer, but they argue that understanding cancer's history could reveal clues that would otherwise remain hidden. "Obviously, we always have that in the back of our minds in everything we do," says Judith Campisi of Lawrence Berkeley National Laboratory.

The Dawn of Cancer

At its root, cancer is a disease of multicellularity. Our single-celled ancestors reproduced by dividing in two. After animals emerged, about seven hundred million years ago, the cells inside their bodies continued to reproduce by dividing, using the molecular machinery they inherited from their progenitors. The cells also began to specialize as they divided, forming different tissues. The complex, multicellular bodies animals have today were made possible by the emergence of new genes that could control how cells divided—such as by stopping the cells' reproduction once an organ reached its adult size. The millions of animal species are evidence of the great evolutionary success that came with acquiring a body. But bodies also present a profound risk. Whenever a cell inside a body divides, its DNA has a small chance of acquiring a cancer-causing mutation. "Every time a cell divides, it's going to be at risk of developing into cancer," Campisi says.

Rare mutations, for instance, may cause a cell to lose restraint and begin to multiply uncontrollably. Other mutations can add to the problem: they may allow deranged cells to invade surrounding tissues and spread through the body. Or they may allow tumor cells to evade the immune system or attract blood vessels that can supply fresh oxygen.

Cancer, in other words, re-creates within our own bodies the evolutionary process that enables animals to adapt to their environment. At the level of organisms, natural selection operates when genetic mutations cause some organisms to have more reproductive success than others; the mutations get "selected" in the sense that they persist and become more common in future generations. In cancer, cells play the role of organisms. Cancer-causing changes to DNA cause some cells to reproduce more effectively than ordinary ones. And even within a single tumor, more adapted cells may outcompete less successful ones. "It's like Darwinian evolution, except that it happens within one organ," explains Natalia Komarova of the University of California, Irvine.

LIMITS TO DEFENSES

Although our bodies may be vulnerable to cancer, they also have many ways to halt it. These strategies probably resulted from natural selection, because mutations that made our ancestors less likely to die of cancer in their prime could have raised their reproductive success. But given the many millions of people who get cancer every year, it is obvious that these defenses have not eradicated the disease.

By studying the evolution of these defenses, biologists are trying to understand why they fall short.

Tumor suppressor proteins are among the most effective defenses against cancer. Studies suggest that some of these proteins prevent cancer by monitoring how a cell reproduces. If the cell multiplies in an abnormal way, the proteins induce it to die or to slip into senescence, a kind of early retirement. The cell survives, but it can no longer divide. Tumor suppressor proteins play a vital role in our survival, but scientists have recently discovered something strange about them: in some respects, we would be better off without them.

Norman E. Sharpless of the University of North Carolina at Chapel Hill genetically engineered mice to study the effect of one of these proteins, called p16 (or, more properly, p16-Ink4a). He and his colleagues created a line of mice that lacked a functional gene for p16 and thus could not produce the protein. In September 2006 the group published three studies on the mice. As expected, the animals were more prone to cancer, which could arise when they were only a year old.

But losing the p16 gene had an upside. When the mice got old, their cells still behaved as if they were young. In one experiment, the scientists studied older mice, some of which had working p16 genes and some of which did not. They destroyed insulin-producing cells in the pancreases of the animals. The normal rodents could no longer produce insulin and developed fatal diabetes. But the ones without the p16 protein developed only mild diabetes and survived. The progenitors of their insulin-producing cells could still multiply quickly, and they repopulated the pancreas with new cells. The scientists found similar results when they examined cells in the blood and brains of the mice: p16 protected them against cancer but also made them old.

These results support a hypothesis Campisi has developed over the past few years. Natural selection favors anticancer proteins such as p16, but only in moderation. If these proteins become too aggressive, they can create their own threats to health by making bodies age too quickly. "It's still a working hypothesis," Campisi admits, "but the data are looking stronger and stronger."

DELAYING THE INEVITABLE

A defense against cancer does not have to eradicate the disease completely to be favored by natural selection. If it can just delay tumors until old age, it may allow people to have more children, on average, than others who lack the defense. It may seem cruel for evolution to stick old people with cancer, but as Jarle Breivik of the University of Oslo points out, "natural selection does not favor genes because they let us live long and happy lives. They are selected for their ability to propagate their information through the generations."

Anticancer proteins such as p16 may favor the young over the old. When p16 pushes a cell into senescence, the cell does not just stop multiplying. It also begins producing an odd balance of proteins. Among the proteins it makes is vascular endothelial growth factor (VEGF), which triggers the growth of more blood vessels. VEGF fosters the growth of tumors by supplying them with extra nutrients. In young people, p16's main effect may be to suppress cancerous cells. But over time, it may create a growing population of senescent cells, which could make people more vulnerable to cancer in old age.

Another way to delay cancer is to set up several lines of defense. Studies on colon cancer, for example, show that cells in the colon must acquire mutations to several genes before they turn cancerous. These defense lines do not prevent people from getting colon cancer—in fact, it is the third most common form of the disease. But the need for multiple mutations to occur in a cell may reduce the chances that colon cancer will arise in young individuals. The average age of people diagnosed with colon cancer is seventy.

Not all cancers strike the old, of course. Most victims of a cancer of the retina called retinoblastoma, for example, are children. But Leonard Nunney of the University of California, Riverside, argues that evolution is responsible for that difference between the two cancers. Nunney points out that colon cells have many more opportunities for acquiring dangerous mutations than retinal cells do. The colon is a large organ made of many cells, which continue replicating throughout a person's life as old cells slough off and new ones take their place. That risk puts a big evolutionary premium on defenses that can prevent colon cells from turning cancerous.

The retina, on the other hand, is "the smallest bit of tissue you can imagine," as Nunney puts it. That small set of retinal cells also stops multiplying by the time a child turns five. With fewer cell divisions occurring, the retina has far fewer opportunities to turn cancerous. As a result, retinoblastoma is extremely rare, striking only four people in a million. Because the risk is so much lower, Nunney argues, natural selection cannot drive the spread of new defenses against retinoblastoma. A defense against cancer in the retina would make very little difference to the average reproductive success of a population.

MAKING TOOLS FOR TUMORS

Recent research suggests that natural selection may have altered genes in ways that make cancer cells more dangerous. Evolutionary biologists discovered this disturbing possibility as they searched for the changes that have made us uniquely human. After our ancestors diverged from other apes about six million years ago, they experienced natural selection as they adapted to a new way of life as a toolmaking, savanna-walking hominid. Scientists can distinguish between genes that have not changed significantly since the origin of hominids and those that have undergone major alteration as a result of selection pressures. It turns out that among the genes that have changed most dramatically are some that play important roles in cancer.

Scientists suspect that the adaptive advantages brought by these genes outweigh the harm they may cause. One of these highly evolved cancer genes makes a protein called fatty acid synthase

(FAS). Normal cells use the protein encoded by this gene to make some of their fatty acids, which are used for many functions, such as building membranes and storing energy. In tumors, however, cancer cells produce FAS protein at a much higher rate. The protein is so important to them that blocking the activity of the gene can kill cancer cells. By comparing the sequence of the FAS gene in humans and other mammals, Mary J. O'Connell of Dublin City University and James McInerney of the National University of Ireland found that the gene has undergone strong natural selection in humans. "This gene has really changed in our lineage," McInerney says.

McInerney cannot say what FAS does differently in humans, but he is intrigued by a hypothesis put forward by the late psychiatrist David Horrobin in the 1990s. Horrobin argued that the dramatic increase in the size and power of the human brain was made possible by the advent of new kinds of fatty acids. Neurons need fatty acids to build membranes and make connections. "One of the things that might allow a larger brain size was our ability to synthesize fats," McInerney speculates. But with that new ability may have come a new tool that cancer cells could borrow for their own ends. Cancer cells may, for example, use FAS as an extra source of energy.

Many fast-evolving cancer genes normally produce proteins in tissues involved in reproduction—in the placenta, for example. Bernard Crespi of Simon Fraser University in British Columbia and Kyle Summers of East Carolina University argue that these genes are part of an evolutionary struggle between children and their mothers.

Natural selection favors genes that allow children to draw as much nourishment from their mothers as possible. A fetus produces the placenta, which grows aggressively into the mother's tissue and extracts nutrients. That demand puts the fetus in conflict with its mother. Natural selection also favors genes that allow mothers to give birth to healthy children. If a mother sacrifices too much in the pregnancy of one child, she may be less likely to have healthy children afterward.

So mothers produce compounds that slow down the flow of nutrients into the fetus.

Each time mothers evolve new strategies to restrain their fetuses, natural selection favors mutations that allow the fetuses to overcome those strategies. "It's a restrained conflict. There's a tug-of-war about how much the fetus is going to take from the mother," Crespi says.

Genes that allow cells to build a better placenta, Crespi and Summers argue, can get hijacked by cancer cells—turned on when they would normally be silent. The ability to stimulate new blood vessel formation and aggressive growth serves a tumor just as it does a placenta. "It's something naturally liable to be co-opted by cancer cell lineages," Summers says. "It sets up the opportunity for mutations to create tools for cancer cells to use to take over the body."

Yet even though activation of these usually quiet genes may make cancers more potent, natural selection may still have favored them because they helped fetuses grow. "You may get selection for a gene variant that helps the fetus get a little more from mom," Crespi says. "But then, when that kid is sixty, it might increase the odds of cancer by a few percent. It's still going to be selected for because of the strong positive early effects."

Sperm are another kind of cell that multiplies rapidly. But whereas placental cells proliferate for a few months, sperm-making cells function for a lifetime. "For decades, human males are producing an enormous amount of sperm all the time," says Andrew Simpson of the Ludwig Institute for Cancer Research in New York City. Genes that operate specifically in such cells are also among the fastest evolving in the human genome. A gene that allows a progenitor sperm cell to divide faster than other cells will become more common in a man's population of sperm. That means it will be more likely to get into a fertilized egg and be passed down to future generations.

Unfortunately for us, genes that make for fast-breeding sperm cells can make for fast-breeding cancer cells. Normally, nonsperm cells prevent these genes from making proteins. "These are genes that need to be firmly silenced, because they are dangerous genes,"

Simpson says. It appears that in cancer cells, mutations can un-
lock these sperm genes, allowing the cells to multiply quickly.

HOW VS. WHY

Evolutionary biologists hope that their research can help in the fight
against cancer. In addition to clarifying why evolution has not eradicated cancer, evolutionary biology may shed light on one of the
most daunting challenges faced by oncologists: the emergence of
drug-resistant tumors.

Chemotherapy drugs often lose their effectiveness against cancer
cells. The process has many parallels to the evolution of resistance to
antiviral drugs in HIV. Mutations that allow cancer cells to survive
exposure to chemotherapy drugs enable the tumor cells to outcompete more vulnerable cells. Understanding the evolution of HIV and
other pathogens has helped scientists to come up with new strategies
for avoiding resistance. Now scientists are investigating how understanding the evolution within tumors could lead to better ways of
using chemotherapy.

The concepts evolutionary biologists have been exploring are relatively new for most cancer biologists. Some are reacting with great enthusiasm. Simpson believes, for instance, that deciphering the rapid
evolution of sperm-related genes could help in the fight against tumors that borrow them. "I think it's absolutely crucial to understand exactly why there is such strong selection on these genes,"
Simpson says. "Understanding that will give us a real insight into
cancer."

Bert Vogelstein of the Howard Hughes Medical Institute also
finds it useful to view cancer through an evolutionary lens. "Thinking about cancer in evolutionary terms jibes perfectly with the views
of cancer molecular geneticists," he says. "In one sense, cancer is a
side effect of evolution."

But Vogelstein is not yet persuaded by the significance of fast-
evolving cancer genes. "One has to be a little cautious. The first ques-

tion I would ask is, Are they looking at the whole genome in a wholly unbiased way?" McInerney acknowledges that such systematic studies have not yet been conducted, but the early results have prompted him and other scientists to begin them.

Some cancer specialists are leery of the entire approach. Christopher Benz of the Buck Institute for Age Research says that any insights from evolution should not be accepted until they are put to an experimental test the way any other hypothesis would be. "Call me skeptical," he says.

Crespi is familiar with this skepticism, and he thinks that it may emerge from the different kinds of questions evolutionary biologists and cancer biologists ask. "The people working on cancer are working on the how question, and the evolution people are working on the why," he says.

Perhaps by asking different questions, evolutionary biologists will be able to contribute to some of the debates among cancer biologists. One long-standing argument focuses on whether mice are good models for cancer in humans. Some evolutionary biologists argue that they are not, because of their separate history. Rodents inherited the same set of genes as we did from our common ancestor some one hundred million years ago, but then many of those genes underwent more change in the two lineages. Cancer-related genes such as FAS may have experienced intense evolutionary change in humans in just the past few million years, making them significantly different from their counterparts in mice.

Mice may also be a poor choice for a cancer model because of the way they reproduce. Scientists have bred lab mice to produce more pups at a faster rate than their wild cousins. Such manipulation may have altered the evolutionary trade-off faced by mice, so that they are rewarded for investing energy into growing quickly and reproducing young.

At the same time, this artificial selection may be selecting against cancer defenses. "We have changed their life histories by selecting on their timing of reproduction," Crespi says.

Ultimately, the study of the evolution of cancer may reveal why eradicating the disease has proved so difficult. "There is no real solution to the problem," Breivik says. "Cancer is a fundamental consequence of the way we are made. We are temporary colonies made by our genes to propagate them to the next generation. The ultimate solution to cancer is that we would have to start reproducing ourselves in a different way."

TARA PARKER-POPE

How NIH Misread Hormone Study in 2002

FROM THE *WALL STREET JOURNAL*

The 2002 study that concluded, explosively, that hormone therapy for menopause led to a greater risk of heart attack turned many women away from that treatment. More recent reviews of the data have pointed to different results, however. Tara Parker-Pope examines what changed.

ON JULY 9, 2002, federal government health officials announced that they had halted a major study of menopause hormones, saying the drugs increased a woman's risk of heart attack by 29 percent.

But in the five years since, it's become clear that some aspects of what was initially reported from the $725 million Women's Health

Initiative study were either misleading or just wrong. Although the government initially said the findings applied to all women, regardless of age or health status, additional data published in recent months show that the age of a woman and the timing of hormone use dramatically changes the risk and benefits. WHI data published in April in the *Journal of the American Medical Association* showed that women in their fifties who took a combination of estrogen and progestin or estrogen alone had a 30 percent lower risk of dying than women who didn't take hormones.

Last month, the *New England Journal of Medicine* reported that fifty- to fifty-nine-year-old women in the WHI who regularly used estrogen alone showed a 60 percent lower risk for severe coronary artery calcium, an important risk factor for heart attack.

How could the heart risks of menopause hormones for this crucial cohort change so dramatically in just five years? Officials from the National Institutes of Health, which directed the study of more than twenty-seven thousand women, say the interpretation of the WHI has simply evolved as researchers have used different methods to analyze the voluminous body of data.

The average age of women in the study was sixty-three. While older women in the study did show a heart risk, researchers eventually focused on women in their fifties who were closer to menopause, finding that hormones were more likely to protect those women's hearts than harm them.

But critics, including some of the WHI's own investigators, speaking out for the first time, say that NIH officials initially overgeneralized in large part because they excluded many of the study's own investigators and physicians from the first review. As a result, key questions that could have clarified the data far sooner weren't asked.

Just eleven days before the public announcement in July 2002, the WHI's forty investigators met in Chicago, where they were told the study had been stopped early. Several people who attended the meeting say several WHI researchers were stunned and angry when they were given final page proofs of the study report for the *Journal of the American Medical Association.* Although some researchers expressed concern that the results were too broadly interpreted, it was too late to make meaningful changes to the *JAMA* article.

Many investigators who had spent nearly a decade working on the WHI had no input in the final and most important paper.

"I think that had the initial report been written by a broader group, as almost all of our later papers have been, it would have been framed differently," says Robert D. Langer, the former principal investigator for the WHI's clinical center at the University of California, San Diego, who was among those who protested at the time. He has since served as an expert witness for hormone maker Wyeth. Dr. Langer says he remains concerned that the interpretation of the WHI has unnecessarily scared a generation of women from the treatment.

Jacques Rossouw, a physician with the National Heart, Lung, and Blood Institute who has overseen the WHI since its inception, confirms that some investigators were upset that they weren't included in writing the first WHI report. "That was an NIH decision supported by the WHI executive committee to keep it to a small group because we realized it was a sensitive paper," he says.

Still, he defends the government's handling of the study results "based on what we knew at the time," and says that study officials wanted to make a dramatic statement. "Our main job at the time was to turn around the prevailing notion that hormones would be useful for long-term prevention of heart disease," he says. "That was our objective. That was a worthy objective which we achieved."

But many in the medical community disagree, saying key questions about long-term use still aren't answered. Although the WHI data clearly show that starting hormones at an older age is risky, what's not clear is whether the heart protection women get starting at a younger age will continue with long-term use.

This was one of the questions that the WHI was supposed to answer when it was launched in 1991 after data from an ongoing trial of nurses showed that women who used menopause hormones have as much as a 50 percent lower risk of heart attack. But the WHI designers didn't take into account that the timing of hormone use might affect the results and recruited mostly older, symptom-free women. Some of the study participants were already twenty years past menopause when the WHI began.

Since mostly older women were recruited, there weren't enough recently menopausal women under sixty to generate conclusive data in some of the findings. But the trends were provocative. Among recently menopausal women who used estrogen and progestin, heart attack risk fell 11 percent. By comparison, women who started taking hormones ten or more years past menopause had a 22 to 71 percent higher risk of heart problems.

In a second part of the study, in which women who had undergone hysterectomy took estrogen without progestin, women who started hormones after the age of 70 had an 11 percent higher risk of heart problems. But women below sixty in the estrogen-only study had a 37 percent lower heart risk.

In a bid to draw a more definitive conclusion, the WHI in April 2007 published a report in *JAMA* combining the data from both hormone trials. That paper showed that the timing of hormone use matters: Younger women appear to receive heart protection, while older women are at risk.

"We now have a refined understanding of the benefits and risks of hormone therapy, and there has been so much reassuring evidence for younger women over the last few years," says Harvard professor and WHI investigator JoAnn Manson, who has no ties to drug firms.

Despite the recent data, questions about long-term use of hormones are far from resolved. While most people now agree that hormones are a reasonable option for women to treat menopause symptoms like hot flashes, the bigger question is whether hormones should be considered in the armamentarium of drugs used for heart protection, along with blood pressure medications and cholesterol-lowering drugs called statins.

The National Heart, Lung, and Blood Institute, which oversaw the WHI, firmly believes that menopause hormones shouldn't be used to prevent heart disease because of potential risk of blood clot, stroke, and breast cancer.

Indeed, the WHI showed that hormones have a range of risks and benefits. On the plus side, they may protect younger women's hearts, they definitely protect against hip fractures, they lower the risk for diabetes, and may lower the risk for colon cancer.

On the other hand, menopause hormones do increase the risk for blood clots and stroke. Women in the WHI who used both estrogen and progestin were at 24 percent higher risk for breast cancer. But women who regularly use estrogen without progestin had a 33 percent lower risk of breast cancer.

"Hormone therapy long term has these other adverse events hanging around. It doesn't fit the paradigm of what you are looking for in a viable long-term prevention strategy," says Dr. Rossouw. "If you're going to use something to prevent atherosclerosis, your choice is statins, not hormones."

The data on women who use statins are mixed, but suggest that they lower a woman's heart risk by 15 percent to 20 percent. Statins carry their own risks, including liver effects, muscle pain, memory problems, and in rare cases a life-threatening muscle disease called rhabdomyolysis. There are no studies showing the risks and benefits of long-term use of statins.

Many doctors now believe that for younger women without a uterus, estrogen should be an option for long-term prevention of heart disease. (The progestin is added to protect against uterine cancer.)

"If this is preventing heart disease and saving lives, I think it's really wrong not to consider it," says Yale associate professor Hugh Taylor, a principal investigator for the Kronos Early Estrogen Prevention Study, a study of estrogen and heart disease funded by

Arizona billionaire John Sperling, an education entrepreneur who was upset with the way the WHI study was initially interpreted. "Some of the other drugs we use for cardiovascular disease don't have the evidence that we have for hormone therapy."

Two important ongoing studies will further illuminate the role estrogen plays in heart disease. The KEEPS study is recruiting 720 women, ages forty-two to fifty-eight, to study the effects of oral or transdermal estrogen as well as progesterone on the coronary arteries of healthy women. An NIH-funded Early Versus Late Intervention Trial will study 500 women and the effects of hormone therapy given within six years of menopause compared to treatment given ten or more years after menopause. Both studies will also test whether using natural progesterone, instead of the synthetic progestin used in the WHI, lowers or eliminates the risk of breast cancer associated with combination therapy.

Even though the long-term heart issue is unresolved, some critics say the NIH's handling of the WHI data scared away younger women who might want to use hormones for menopause symptoms. "We've gone to considerable efforts to reassure women and gynecologists that it's OK [to use hormones] in the short term," to treat the symptoms of menopause, Dr. Rossouw says.

Still, since the WHI results were announced in 2002, hormone sales have plummeted 30 percent to $1.9 billion, according to IMS Health, a health care information company.

"I don't think it's fair to extrapolate from the data that women should be put on this as a preventative treatment for heart disease," says Michael E. Mendelsohn, a Tufts University professor who recently wrote a *NEJM* editorial about estrogen and heart disease. "What the data do support is that women who use hormones to treat menopause symptoms can feel reassured that they are not increasing their cardiovascular risk and may be providing some long-term benefit."

GARDINER HARRIS, BENEDICT CAREY, AND JANET ROBERTS

Psychiatrists, Children, and Drug Industry's Role

FROM THE *NEW YORK TIMES*

> *In their investigation of how pharmaceutical companies may be influencing doctors, Gardiner Harris, Benedict Carey, and Janet Roberts focus on psychiatry—a field especially vulnerable to the dangers of "atypical" prescribing, where medications are prescribed for disorders they were not originally approved to treat.*

WHEN ANYA BAILEY DEVELOPED an eating disorder after her twelfth birthday, her mother took her to a psychiatrist at the University of Minnesota who prescribed a powerful antipsychotic drug called Risperdal.

Created for schizophrenia, Risperdal is not approved to treat eating disorders, but increased appetite is a common side effect and doctors may prescribe drugs as they see fit. Anya gained weight but within two years developed a crippling knot in her back. She now receives regular injections of Botox to unclench her back muscles. She often awakens crying in pain.

Isabella Bailey, Anya's mother, said she had no idea that children might be especially susceptible to Risperdal's side effects. Nor did she know that Risperdal and similar medicines were not approved at the time to treat children, or that medical trials often cited to justify the use of such drugs had as few as eight children taking the drug by the end.

Just as surprising, Ms. Bailey said, was learning that the university psychiatrist who supervised Anya's care received more than seven thousand dollars from 2003 to 2004 from Johnson & Johnson, Risperdal's maker, in return for lectures about one of the company's drugs.

Doctors, including Anya Bailey's, maintain that payments from drug companies do not influence what they prescribe for patients.

But the intersection of money and medicine, and its effect on the well-being of patients, has become one of the most contentious issues in health care. Nowhere is that more true than in psychiatry, where increasing payments to doctors have coincided with the growing use in children of a relatively new class of drugs known as atypical antipsychotics.

These bestselling drugs, including Risperdal, Seroquel, Zyprexa, Abilify, and Geodon, are now being prescribed to more than half a million children in the United States to help parents deal with behavior problems despite profound risks and almost no approved uses for minors.

A *New York Times* analysis of records in Minnesota, the only state that requires public reports of all drug company marketing payments to doctors, provides rare documentation of how financial relationships between doctors and drug makers correspond to the growing use of atypicals in children.

From 2000 to 2005, drug maker payments to Minnesota psychiatrists rose more than sixfold, to $1.6 million. During those same years, prescriptions of antipsychotics for children in Minnesota's Medicaid program rose more than ninefold.

Those who took the most money from makers of atypicals tended to prescribe the drugs to children the most often, the data suggest. On average, Minnesota psychiatrists who received at least five thousand dollars from atypical makers from 2000 to 2005 appear to have written three times as many atypical prescriptions for children as psychiatrists who received less or no money.

The *Times* analysis focused on prescriptions written for about one-third of Minnesota's Medicaid population, almost all of whom are disabled. Some doctors were misidentified by pharmacists, but the information provides a rough guide to prescribing patterns in the state.

Drug makers underwrite decision makers at every level of care. They pay doctors who prescribe and recommend drugs, teach about the underlying diseases, perform studies, and write guidelines that other doctors often feel bound to follow.

But studies present strong evidence that financial interests can affect decisions, often without people knowing it.

In Minnesota, psychiatrists collected more money from drug makers from 2000 to 2005 than doctors in any other specialty. Total payments to individual psychiatrists ranged from $51 to more than $689,000, with a median of $1,750. Since the records are incomplete, these figures probably underestimate doctors' actual incomes.

Such payments could encourage psychiatrists to use drugs in ways that endanger patients' physical health, said Dr. Steven E. Hyman, the provost of Harvard University and former director of the National Institute of Mental Health. The growing use of atypicals in children is the most troubling example of this, Dr. Hyman said.

"There's an irony that psychiatrists ask patients to have insights into themselves, but we don't connect the wires in our own lives about how money is affecting our profession and putting our patients at risk," he said.

THE PRESCRIPTION

Anya Bailey is a fifteen-year-old high school freshman from East Grand Forks, Minnesota, with pictures of the actor Chad Michael Murray on her bedroom wall. She has constant discomfort in her neck that leads her to twist it in a birdlike fashion. Last year, a boy mimicked her in the lunch room.

"The first time, I laughed it off," Anya said. "I said: 'That's so funny. I think I'll laugh with you.' Then it got annoying, and I decided to hide it. I don't want to be made fun of."

Now she slumps when seated at school to pressure her clenched muscles, she said.

It all began in 2003 when Anya became dangerously thin. "Nothing tasted good to her," Ms. Bailey said.

Psychiatrists at the University of Minnesota, overseen by Dr. George M. Realmuto, settled on Risperdal, not for its calming effects but for its normally unwelcome side effect of increasing appetite and weight gain, Ms. Bailey said. Anya had other issues that may have recommended Risperdal to doctors, including occasional angry outbursts and having twice heard voices over the previous five years, Ms. Bailey said.

Dr. Realmuto said he did not remember Anya's case, but speaking generally he defended his unapproved use of Risperdal to counter an eating disorder despite the drug's risks. "When things are dangerous, you use extraordinary measures," he said.

Ten years ago, Dr. Realmuto helped conduct a study of Concerta, an attention deficit hyperactivity disorder drug marketed by Johnson & Johnson, which also makes Risperdal. When Concerta was approved, the company hired him to lecture about it.

He said he gives marketing lectures for several reasons.

"To the extent that a drug is useful, I want to be seen as a leader in my specialty and that I was involved in a scientific study," he said.

The money is nice, too, he said. Dr. Realmuto's university salary is $196,310.

"Academics don't get paid very much," he said. "If I was an entertainer, I think I would certainly do a lot better."

In 2003, the year Anya came to his clinic, Dr. Realmuto earned five thousand dollars from Johnson & Johnson for giving three talks about Concerta. Dr. Realmuto said he could understand someone's worrying that his Concerta lecture fees would influence him to prescribe Concerta but not a different drug from the same company, like Risperdal.

In general, he conceded, his relationship with a drug company might prompt him to try a drug. Whether he continued to use it, though, would depend entirely on the results.

As the interview continued, Dr. Realmuto said that upon reflection his payments from drug companies had probably opened his door to useless visits from a drug salesman, and he said he would stop giving sponsored lectures in the future.

Kara Russell, a Johnson & Johnson spokeswoman, said that the company selects speakers who have used the drug in patients and have either undertaken research or are aware of the studies. "Dr. Realmuto met these criteria," Ms. Russell said.

When asked whether these payments may influence doctors' prescribing habits, Ms. Russell said that the talks "provide an educational opportunity for physicians."

No one has proved that psychiatrists prescribe atypicals to children because of drug company payments. Indeed, some who frequently prescribe the drugs to children earn no drug industry money. And nearly all psychiatrists who accept payments say they remain independent. Some say they prescribed and extolled the benefits of such drugs before ever receiving payments to speak to other doctors about them.

"If someone takes the point of view that your doctor can be bought, why would you go to an ER with your injured child and say, 'Can you help me?'" said Dr. Suzanne A. Albrecht, a psychiatrist from Edina, Minnesota, who earned more than $188,000 from 2002 to 2005 giving drug marketing talks.

THE INDUSTRY CAMPAIGN

It is illegal for drug makers to pay doctors directly to prescribe specific products. Federal rules also bar manufacturers from promoting unapproved, or off-label, uses for drugs.

But doctors are free to prescribe as they see fit, and drug companies can sidestep marketing prohibitions by paying doctors to give lectures in which, if asked, they may discuss unapproved uses.

The drug industry and many doctors say that these promotional lectures provide the field with invaluable education. Critics say the payments and lectures, often at expensive restaurants, are disguised kickbacks that encourage potentially dangerous drug uses. The issue is particularly important in psychiatry, because mental problems are not well understood, treatment often involves trial and error, and off-label prescribing is common.

The analysis of Minnesota records shows that from 1997 through 2005, more than a third of Minnesota's licensed psychiatrists took money from drug makers, including the last eight presidents of the Minnesota Psychiatric Society.

The psychiatrist receiving the most from drug companies was Dr. Annette M. Smick, who lives outside Rochester, Minnesota, and was paid more than $689,000 by drug makers from 1998 to 2004. At one point Dr. Smick was doing so many sponsored talks that "it was hard for me to find time to see patients in my clinical practice," she said.

"I was providing an educational benefit, and I like teaching," Dr. Smick said.

Dr. Steven S. Sharfstein, immediate past president of the American Psychiatric Association, said psychiatrists have become too cozy with drug makers. One example of this, he said, involves Lexapro, made by Forest Laboratories, which is now the most widely used antidepressant in the country even though there are cheaper alternatives, including generic versions of Prozac.

"Prozac is just as good if not better, and yet we are migrating to the expensive drug instead of the generics," Dr. Sharfstein said. "I think it's the marketing."

Atypicals have become popular because they can settle almost any extreme behavior, often in minutes, and doctors have few other answers for desperate families.

Their growing use in children is closely tied to the increasingly common and controversial diagnosis of pediatric bipolar disorder, a mood problem marked by aggravation, euphoria, depression, and, in some cases, violent outbursts. The drugs, sometimes called major tranquilizers, act by numbing brain cells to surges of dopamine, a chemical that has been linked to euphoria and psychotic delusions.

Suzette Scheele of Burnsville, Minnesota, said her seventeen-year-old son, Matt, was given a diagnosis of bipolar disorder four years ago because of intense mood swings, and now takes Seroquel and Abilify, which have caused substantial weight gain.

"But I don't have to worry about his rages; he's appropriate; he's pleasant to be around," Ms. Scheele said.

The sudden popularity of pediatric bipolar diagnosis has coincided with a shift from antidepressants like Prozac to far more expensive atypicals. In 2000, Minnesota spent more than $521,000 buying antipsychotic drugs, most of it on atypicals, for children on Medicaid. In 2005, the cost was more than $7.1 million, a fourteen-fold increase.

The drugs, which can cost one thousand to eight thousand dollars for a year's supply, are huge sellers worldwide. In 2006, Zyprexa, made by Eli Lilly, had $4.36 billion in sales, Risperdal $4.18 billion, and Seroquel, made by AstraZeneca, $3.42 billion.

Many Minnesota doctors, including the president of the Minnesota

Psychiatric Society, said drug makers and their intermediaries are now paying them almost exclusively to talk about bipolar disorder.

The Diagnoses

Yet childhood bipolar disorder is an increasingly controversial diagnosis. Even doctors who believe it is common disagree about its telltale symptoms. Others suspect it is a fad. And the scientific evidence that atypicals improve these children's lives is scarce.

One of the first and perhaps most influential studies was financed by AstraZeneca and performed by Dr. Melissa DelBello, a child and adult psychiatrist at the University of Cincinnati.

Dr. DelBello led a research team that tracked for six weeks the moods of thirty adolescents who had received diagnoses of bipolar disorder. Half of the teenagers took Depakote, an antiseizure drug used to treat epilepsy and bipolar disorder in adults. The other half took Seroquel and Depakote.

The two groups did about equally well until the last few days of the study, when those in the Seroquel group scored lower on a standard measure of mania. By then, almost half of the teenagers getting Seroquel had dropped out because they missed appointments or the drugs did not work. Just eight of them completed the trial.

In an interview, Dr. DelBello acknowledged that the study was not conclusive. In the 2002 published paper, however, she and her coauthors reported that Seroquel in combination with Depakote "is more effective for the treatment of adolescent bipolar mania" than Depakote alone.

In 2005, a committee of prominent experts from across the country examined all of the studies of treatment for pediatric bipolar disorder and decided that Dr. DelBello's was the only study involving atypicals in bipolar children that deserved its highest rating for scientific rigor. The panel concluded that doctors should consider atypicals as a first-line treatment for some children. The guidelines were published in the *Journal of the American Academy of Child and Adolescent Psychiatry*.

Three of the four doctors on the panel served as speakers or consultants to makers of atypicals, according to disclosures in the guidelines. In an interview, Dr. Robert A. Kowatch, a psychiatrist at Cincinnati Children's Hospital and the lead author of the guidelines, said the drug makers' support had no influence on the conclusions.

AstraZeneca hired Dr. DelBello and Dr. Kowatch to give sponsored talks. They later undertook another study comparing Seroquel and Depakote in bipolar children and found no difference. Dr. DelBello, who earns $183,500 annually from the University of Cincinnati, would not discuss how much she is paid by AstraZeneca.

"Trust me, I don't make much," she said. Drug company payments did not affect her study or her talks, she said. In a recent disclosure, Dr. DelBello said that she received marketing or consulting income from eight drug companies, including all five makers of atypicals.

Dr. Realmuto has heard Dr. DelBello speak several times, and her talks persuaded him to use combinations of Depakote and atypicals in bipolar children, he said. "She's the leader in terms of doing studies on bipolar," Dr. Realmuto said.

Some psychiatrists who advocate use of atypicals in children acknowledge that the evidence supporting this use is thin. But they say children should not go untreated simply because scientists have failed to confirm what clinicians already know.

"We don't have time to wait for them to prove us right," said Dr. Kent G. Brockmann, a psychiatrist from the Twin Cities who made more than sixteen thousand dollars from 2003 to 2005 doing drug talks and one-on-one sales meetings, and last year was a leading prescriber of atypicals to Medicaid children.

THE REACTION

For Anya Bailey, treatment with an atypical helped her regain her appetite and put on weight, but also heavily sedated her, her mother said. She developed the disabling knot in her back, the result of a nerve condition called dystonia, in 2005.

The reaction was rare but not unknown. Atypicals have side effects that are not easy to predict in any one patient. These include rapid weight gain and blood sugar problems, both risk factors for diabetes; disfiguring tics; dystonia; and in rare cases heart attacks and sudden death in the elderly.

In 2006, the Food and Drug Administration received reports of at least 29 children dying and at least 165 more suffering serious side effects in which an antipsychotic was listed as the "primary suspect." That was a substantial jump from 2000, when there were at least 10 deaths and 85 serious side effects among children linked to the drugs. Since reporting of bad drug effects is mostly voluntary, these numbers likely represent a fraction of the toll.

Jim Minnick, a spokesman for AstraZeneca, said that the company carefully monitors reported problems with Seroquel. "AstraZeneca believes that Seroquel is safe," Mr. Minnick said.

Other psychiatrists renewed Anya's prescriptions for Risperdal until Ms. Bailey took Anya last year to the Mayo Clinic, where a doctor insisted that Ms. Bailey stop the drug. Unlike most universities and hospitals, the Mayo Clinic restricts doctors from giving drug marketing lectures.

Ms. Bailey said she wished she had waited to see whether counseling would help Anya before trying drugs. Anya's weight is now normal without the help of drugs, and her counseling ended in March. An experimental drug, her mother said, has recently helped the pain in her back.

GARDINER HARRIS AND JANET ROBERTS

After Sanctions, Doctors Get Drug Company Pay

FROM THE *NEW YORK TIMES*

Pharmaceutical companies are paying disgraced doctors, censured by medical boards, to conduct clinical trials or perform as speakers. In an investigation for the New York Times, *Gardiner Harris and Janet Roberts take a closer look at this questionable practice.*

A DECADE AGO THE MINNESOTA BOARD of Medical Practice accused Dr. Faruk Abuzzahab of a "reckless, if not willful, disregard" for the welfare of forty-six patients, five of whom died in his care or shortly afterward. The board suspended his license for seven months and restricted it for two years after that.

But Dr. Abuzzahab, a Minneapolis psychiatrist, is still overseeing the testing of drugs on patients and is being paid by pharmaceutical companies for the work. At least a dozen have paid him for research or marketing since he was disciplined.

Medical ethicists have long argued that doctors who give experimental medicines should be chosen with care. Indeed, the drug industry's own guidelines for clinical trials state, "Investigators are selected based on qualifications, training, research or clinical expertise in relevant fields." Yet Dr. Abuzzahab is far from the only doctor to have been disciplined or criticized by a medical board but later paid by drug makers.

An analysis of state records by the *New York Times* found more than a hundred such doctors in Minnesota, at least two with criminal fraud convictions. While Minnesota is the only state to make its records publicly available, the problem, experts say, is national.

One of Dr. Abuzzahab's patients was David Olson, whom the psychiatrist tried repeatedly to recruit for clinical trials. Drug makers paid Dr. Abuzzahab thousands of dollars for every patient he recruited. In July 1997, when Mr. Olson again refused to be a test subject, Dr. Abuzzahab discharged him from the hospital even though he was suicidal, records show. Mr. Olson committed suicide two weeks later.

In its disciplinary action against Dr. Abuzzahab, the state medical board referred to Mr. Olson as Patient No. 46.

"Dr. Abuzzahab failed to appreciate the risks of taking Patient No. 46 off Clozaril, failed to respond appropriately to the patient's rapid deterioration and virtually ignored this patient's suicidality," the board found.

In an interview, Dr. Abuzzahab dismissed the findings as "without heft" and said drug makers were aware of his record. He said he had helped study many of the most popular drugs in psychiatry, including Paxil, Prozac, Risperdal, Seroquel, Zoloft, and Zyprexa.

The *Times*'s examination of Minnesota's trove of records on drug company payments to doctors found that from 1997 to 2005, at least

103 doctors who had been disciplined or criticized by the state medical board received a total of $1.7 million from drug makers. The median payment over that period was $1,250; the largest was $479,000.

The sanctions by the board ranged from reprimands to demands for retraining to suspension of licenses. Of those 103 doctors, 39 had been penalized for inappropriate prescribing practices, 21 for substance abuse, 12 for substandard care, and 3 for mismanagement of drug studies. A few cases received national news media coverage, but drug makers hired the doctors anyway.

The *Times* included in its analysis any doctor who received drug company payments within ten years of being under medical board sanction. At least 38 doctors received a combined $140,000 while they were still under sanction. Dr. Abuzzahab received more than $55,000 from 1997 to 2005.

Drug makers refused to comment, said they relied on doctors to report disciplinary or criminal cases, or said they were considering changing their hiring systems.

Asked about the Minnesota analysis, the deputy commissioner and chief medical officer of the Food and Drug Administration, Dr. Janet Woodcock, said the federal government needed to overhaul regulations governing clinical trials and the doctors who oversaw them.

"We recognize that we need to modernize the FDA approach in keeping people safe in clinical trials," Dr. Woodcock said.

Drug makers are not required to inform the agency when they discover that investigators are falsifying data, and indeed some have failed to do so in the past. The FDA plans to require such disclosures, Dr. Woodcock said. The agency inspects at most 1 percent of all clinical trials, she said.

Karl Uhlendorf, a spokesman for the Pharmaceutical Research and Manufacturers of America, said the trade group would not comment on the *Times*'s findings.

The records most likely understate the extent of the problem because they are incomplete. And the Minnesota Board of Medical

Practice disciplines a smaller share of the state's doctors than almost any other medical board in the country, according to rankings by Public Citizen, an advocacy group based in Washington.

Dr. David Rothman, president of the Institute on Medicine as a Profession at Columbia University, said the *Times* analysis revealed a national problem. "There's no reason to think Minnesota is unique," Dr. Rothman said.

"Clinical trial investigators must be culled from only the finest physicians in the country," he said, "since they work on the frontiers of new knowledge. That drug makers are scraping the bottom of the medical barrel is an outrage."

Payments by drug companies to doctors, whether or not the doctors have been disciplined, are a matter of much debate. Drug makers and doctors say the money finances vital research and helps educate doctors about helpful medicines. But others in the medical profession say the payments are thinly disguised incentives for doctors to prescribe more, and more expensive, drugs.

Among the other doctors who were disciplined or criticized by the board and paid by pharmaceutical companies:

Dr. Barry Garfinkel, a child psychiatrist from Minneapolis who was convicted in federal court in 1993 of fraud involving a study for Ciba-Geigy. His criminal case made headlines across the state. From 2002 to 2004, Eli Lilly paid him more than $5,500 in honoraria, according to state records.

Dr. Garfinkel said in an interview that he had wondered why drug makers would hire him as a speaker considering his statewide notoriety. He decided that "they're hiring me to influence my prescribing habits," so he quit giving sponsored talks and taking money from drug makers, he said.

Dr. John Simon, a Minneapolis psychiatrist who for years shared an office with Dr. Abuzzahab and was told by the state medical board in 1994 to complete a clinical training program after it concluded in a report that he "frequently makes abrupt and drastic changes in type and dosage of medication which seem erratic, not well considered and poorly integrated with nonmedication strategies." He prescribed addictive drugs to addicts and failed to stop giving medicines to patients suffering severe drug side effects, the board concluded.

Dr. Simon earned more than $350,000 from five drug makers from 1998 to 2005 for consulting and giving drug marketing talks. Of this, Eli Lilly paid more than $314,000. Dr. Simon said in an interview that the board's action was a learning experience, and that drug makers continued to hire him to speak because "I am respected by my peers." Asked about Drs. Garfinkel and Simon, Phil Belt, a spokesman for Eli Lilly, said that both doctors were licensed to practice medicine and that the company relied on doctors to report disciplinary actions or criminal convictions against them.

Dr. Ronald Hardrict, a psychiatrist from Minneapolis who pleaded guilty in 2003 to Medicaid fraud. In 2004 and 2005, he collected more than $63,000 in marketing payments from seven drug makers. In an interview, Dr. Hardrict said it was "insulting" and "ridiculous" to suggest that income from drug makers might influence doctors' prescribing habits.

"I bought the Mercedes because it has air bags, and I use Risperdal because it works," Dr. Hardrict said, referring to an antipsychotic medicine for schizophrenia. Johnson & Johnson, the maker of Risperdal, paid Dr. Hardrict more than $30,000 in 2003 and 2004.

Srikant Ramaswami, a spokesman for Johnson & Johnson, said the company removed Dr. Hardrict as a speaker in 2004 when, as a result of his conviction, his name appeared in a government database.

Asked why other drug makers continue to hire him despite

a fraud conviction, Dr. Hardrict responded with an e-mail message stating only, "I will pray for you daily."

IN CASES INVOLVING DR. ABUZZAHAB over fifteen years in the 1980s and '90s, the medical board found that he repeatedly prescribed narcotics and other controlled substances to addicts, renewing one patient's prescriptions six weeks after the patient was jailed and telling another that his addictive pills should be thought of as "Hamburger Helper." He prescribed narcotics to pregnant patients, one of whom prematurely delivered a baby who soon died.

In explaining his abrupt discharge of the suicidal Mr. Olson, Dr. Abuzzahab told the medical board that "if a patient is determined to kill himself, he can't be prevented from doing it and hospitalization postpones the event," records show.

Mr. Olson's sister, Susie Olson, said Dr. Abuzzahab "had no time for my brother unless David agreed to get into a drug study. He said, 'You're wasting my time and the hospital's.' It was all about money."

Separately, the FDA in 1979 and 1984 concluded that Dr. Abuzzahab had violated the protocols of every study he led that they audited, and reported inaccurate data to drug makers. He routinely oversaw four to eight drug trials simultaneously, often moved patients from one study to another, sometimes gave experimental medicines to patients at their first consultation, and once hospitalized a patient for the sole purpose of enrolling him in a study, the FDA found.

Dr. Abuzzahab, 74, was president of the Minnesota Psychiatric Society and two decades ago was chairman of its continuing education and ethics committees. He would not discuss the specifics of his disciplinary record, saying he did not have the time. But in 1998 he signed an agreement with the board saying that his conduct "constitutes a reasonable basis in law and fact to justify the disciplinary action."

A simple Google search reveals Dr. Abuzzahab's 1998 medical board disciplinary file, which was reported at the time by a local newspaper and a TV station. In 1998, the *Boston Globe* featured Dr. Abuzzahab in a front-page article questioning the safety of psychiatric drug experiments. And in 1999, the NBC program *Dateline* did a segment about a woman who committed suicide while in a drug experiment he supervised.

In June 2006, the medical board criticized Dr. Abuzzahab, this time for writing narcotics prescriptions for patients he knew were using false names, a violation of federal narcotics laws.

Despite all this, drug makers continued to hire him. Dr. Abuzzahab's résumé lists eleven publications or research presentations since 2000, when the medical board lifted its restrictions on his license.

Takeda, a Japanese drug maker, confirmed that Dr. Abuzzahab was doing a study financed by the company on its sleep medicine, Rozerem. Eisai, another Japanese drug maker, said that although Dr. Abuzzahab had signed a clinical trial agreement with the company to study its Alzheimer's drug, Aricept, it told him two days after a reporter asked for comment on the case that he was not qualified to be an investigator. And at AstraZeneca, for which Dr. Abuzzahab said he had performed clinical trials and still gave drug marketing lectures, a spokesman said the company was "concerned" about Dr. Abuzzahab's disciplinary record.

"We have our own internal processes for dealing with these matters, which are under way," said Jim Minnick, an AstraZeneca spokesman.

The Minnesota records often fail to distinguish between drug company payments to doctors for research and for marketing, so it is sometimes impossible to determine why doctors were paid. Some doctors, like Dr. Abuzzahab, clearly performed both research and marketing.

Gene Carbona, who left Merck on good terms in 2001 as a regional sales manager after twelve years in drug sales, said the only thing the company considered when hiring doctors to give marketing lectures was "the volume or potential volume of prescribing that doctor could do."

A Merck spokesman declined to comment.

Mr. Carbona, now executive director of sales for *The Medical Letter*, which reviews drugs, said that had he known that a doctor had a disciplinary record for excessive prescribing, "I would have been more inclined to use them as a speaker."

Daniel Carlat

Dr. Drug Rep

FROM THE *NEW YORK TIMES MAGAZINE*

According to one recent study, some 25 percent of all American doctors receive payment from pharmaceutical companies to lecture to physicians or market drugs in other ways. Daniel Carlat, a psychiatrist, recounts his brief career as a shill for the drug companies.

I. Faculty Development

On a blustery fall New England day in 2001, a friendly representative from Wyeth Pharmaceuticals came into my office in Newburyport, Massachusetts, and made me an offer I found hard to refuse. He asked me if I'd like to give talks to other doctors about using Effexor XR for treating depression. He told me that I would go around to doctors' offices during lunchtime and talk about some of the features of Effexor. It would be pretty easy. Wyeth would provide a set of slides and even pay for me to attend a speaker's training session, and he quickly floated some numbers. I would be paid $500 for one-hour "Lunch and Learn" talks at local doctors' offices, or $750 if I had to drive an hour. I would be flown to New York for a "faculty-development program," where I would be pampered in a Midtown hotel for two nights and would be paid an additional "honorarium."

I thought about his proposition. I had a busy private practice in psychiatry, specializing in psychopharmacology. I was quite familiar with Effexor, since I had read recent studies showing that it might be slightly more effective than SSRIs, the most commonly prescribed antidepressants: the Prozacs, Paxils, and Zolofts of the world. SSRI stands for selective serotonin reuptake inhibitor, referring to the fact that these drugs increase levels of the neurotransmitter serotonin, a chemical in the brain involved in regulating moods. Effexor, on the other hand, was being marketed as a dual reuptake inhibitor, meaning that it increases both serotonin and norepinephrine, another neurotransmitter. The theory promoted by Wyeth was that two neurotransmitters are better than one, and that Effexor was more powerful and effective than SSRIs.

I had already prescribed Effexor to several patients, and it seemed to work as well as the SSRIs. If I gave talks to primary-care doctors about Effexor, I reasoned, I would be doing nothing unethical. It was a perfectly effective treatment option, with some data to suggest advantages over its competitors. The Wyeth rep was simply suggesting that I discuss some of the data with other doctors. Sure, Wyeth would benefit, but so would other doctors, who would become more educated about a good medication.

A few weeks later, my wife and I walked through the luxurious lobby of the Millennium Hotel in Midtown Manhattan. At the reception desk, when I gave my name, the attendant keyed it into the computer and said, with a dazzling smile: "Hello, Dr. Carlat, I see that you are with the Wyeth conference. Here are your materials."

She handed me a folder containing the schedule of talks, an invitation to various dinners and receptions, and two tickets to a Broadway musical. "Enjoy your stay, doctor." I had no doubt that I would, though I felt a gnawing at the edge of my conscience. This seemed like a lot of money to lavish on me just so that I could provide some education to primary-care doctors in a small town north of Boston.

The next morning, the conference began. There were a hundred or so other psychiatrists from different parts of the U.S. I recognized a couple of the attendees, including an acquaintance I hadn't seen in a while. I'd heard that he moved to another state and was making a bundle of money, but nobody seemed to know exactly how.

I joined him at his table and asked him what he had been up to. He said he had a busy private practice and had given a lot of talks for Warner-Lambert, a company that had since been acquired by Pfizer. His talks were on Neurontin, a drug that was approved for epilepsy but that my friend had found helpful for bipolar disorder in his practice. (In 2004, Warner-Lambert pleaded guilty to illegally marketing Neurontin for unapproved uses. It is illegal for companies to pay doctors to promote so-called off-label uses.)

I knew about Neurontin and had prescribed it occasionally for bipolar disorder in my practice, though I had never found it very helpful. A recent study found that it worked no better than a placebo for this condition. I asked him if he really thought Neurontin worked for bipolar, and he said that he felt it was "great for some patients" and that he used it "all the time." Given my clinical experiences with the drug, I wondered whether his positive opinion had been influenced by the money he was paid to give talks.

But I put those questions aside as we gulped down our coffees and took seats in a large lecture room. On the agenda were talks from some of the most esteemed academics in the field, authors of hundreds of articles in the major psychiatric journals. They included Michael Thase, of the University of Pittsburgh and the researcher who single-handedly put Effexor on the map with a meta-analysis, and Norman Sussman, a professor of psychiatry at New York University, who was master of ceremonies.

Thase strode to the lectern first in order to describe his groundbreaking work synthesizing data from more than two thousand patients who had been enrolled in studies comparing Effexor with SSRIs. At this time, with his Effexor study a topic of conversation in the mental-health world, Thase was one of the most well-known and well-respected psychiatrists in the United States. He cut a captivating figure onstage: tall and slim, dynamic, incredibly articulate and a master of the research craft.

He began by reviewing the results of the meta-analysis that had the psychiatric world abuzz. After carefully pooling and processing data from eight separate clinical trials, Thase published a truly significant finding: Effexor caused a 45 percent remission rate in patients in contrast to the SSRI rate of 35 percent and the placebo rate of 25 percent. It was the first time one antidepressant was shown to be more effective than any other. Previously, psychiatrists chose antidepressants based on a combination of guesswork, gut feeling, and tailoring a drug's side effects to a patient's symptom profile. If Effexor was truly more effective than SSRIs, it would amount to a revolution in psychiatric practice and a potential windfall for Wyeth.

One impressive aspect of Thase's presentation was that he was not content to rest on his laurels; rather he raised a series of potential criticisms of his results and then rebutted them convincingly. For example, skeptics had pointed out that Thase was a paid consultant to Wyeth and that both of his coauthors were employees of the company. Thase responded that he had requested and had received all of the company's data and had not cherry-picked from those studies most favorable for Effexor. This was a significant point, because companies sometimes withhold negative data from publication in medical journals. For example, in 2004, GlaxoSmithKline was sued by Eliot Spitzer, who was then the New York attorney general, for suppressing data hinting that Paxil causes suicidal thoughts in children. The company settled the case and agreed to make clinical-trial results public.

Another objection was that while the study was billed as comparing Effexor with SSRIs in general, in fact most of the data compared Effexor with one specific SSRI: Prozac. Perhaps Effexor was, indeed, more effective than Prozac; this did not necessarily mean that it was more effective than the other SSRIs in common use. But Thase announced that since the original study, he had analyzed data on Paxil and other meds and also found differences in remission rates.

For his study, Thase chose what was at that time an unusual measure of antidepressant improvement: "remission," rather than the more standard measure, "response." In clinical antidepressant trials, a "response" is defined as a 50 percent improvement in depressive symptoms, as measured by the Hamilton depression scale. Thus, if a patient enters a study scoring a twenty-four on the Hamilton (which would be a moderate degree of depression), he or she would have "responded" if the final score, after treatment, was twelve or less.

Remission, on the other hand, is defined as "complete" recovery. While you might think that a patient would have to score a zero on the Hamilton to be in remission, in fact very few people score that low, no matter how deliriously happy they are. Instead, researchers come up with various cutoff scores for remission. Thase chose a cutoff score of seven or below.

In his study, he emphasized the remission rates and not the response rates. As I listened to his presentation, I wondered why. Was it because he felt that remission was the only really meaningful outcome by which to compare drugs? Or was it because using remission made Effexor look more impressive than response did? Thase indirectly addressed this issue in his paper by pointing out that even when remission was defined in different ways, with different cutoff points, Effexor beat the SSRIs every time. That struck me as a pretty convincing endorsement of Wyeth's antidepressant.

The next speaker, Norm Sussman, took the baton from Thase and explored the concept of remission in more detail. Sussman's job was to systematically go through the officially sanctioned "slide deck"—slides provided to us by Wyeth, which we were expected to use during our own presentations.

If Thase was the riveting academic, Sussman was the engaging populist, translating some of the drier research concepts into terms that our primary-care-physician audiences would understand. Sussman exhorted us not to be satisfied with response and encouraged us to set the bar higher. "Is the patient doing everything they were doing before they got depressed?" he asked. "Are they doing it even better? That's remission." To further persuade us, he highlighted a slide showing that patients who made it all the way to remission are less likely to relapse to another depressive episode than patients who merely responded. And for all its methodological limitations, it was a slide that I would become well acquainted with, as I would use it over and over again in my own talks.

When it came to side effects, Effexor's greatest liability was that it could cause hypertension, a side effect not shared by SSRIs. Sussman showed us some data from the clinical trials, indicating that at lower doses, about 3 percent of patients taking Effexor had hypertension as compared with about 2 percent of patients assigned to a placebo. There was only a 1 percent difference between Effexor and placebo, he commented, and pointed out that treating high blood pressure might be a small price to pay for relief from depression.

It was an accurate reading of the data, and I remember finding it a convincing defense of Effexor's safety. As I look back at my notes now, however, I notice that another way of describing the same numbers would have been to say that Effexor leads to a 50 percent greater rate of hypertension than a placebo. Framed this way, Effexor looks more hazardous.

And so it went for the rest of the afternoon.

Was I swallowing the message whole? Certainly not. I knew that this was hardly impartial medical education, and that we were being fed a marketing line. But when you are treated like the anointed, wined and dined in Manhattan and placed among the leaders of the field, you inevitably put some of your critical faculties on hold. I was truly impressed with Effexor's remission numbers, and like any physician, I was hopeful that something new and different had been introduced to my quiver of therapeutic options.

At the end of the last lecture, we were all handed envelopes as we left the conference room. Inside were checks for $750. It was time to enjoy ourselves in the city.

II. THE ART AND SCIENCE OF DETAILING

Pharmaceutical "detailing" is the term used to describe those sales visits in which drug reps go to doctors' offices to describe the benefits of a specific drug. Once I returned to my Newburyport office from New York, a couple of voice-mail messages from local Wyeth reps were already waiting for me, inviting me to give some presentations at local doctors' offices. I was about to begin my speaking—and detailing—career in earnest.

How many doctors speak for drug companies? We don't know for sure, but one recent study indicates that at least 25 percent of all doctors in the United States receive drug money for lecturing to physicians or for helping to market drugs in other ways. This meant that I was about to join some two hundred thousand American physicians who are being paid by companies to promote their drugs. I felt quite flattered to have been recruited, and I assumed that the rep had picked me because of some special personal or professional quality.

The first talk I gave brought me back to earth rather quickly. I distinctly remember the awkwardness of walking into my first waiting room. The receptionist slid the glass partition open and asked if I had an appointment.

"Actually, I'm here to meet with the doctor."

"Oh, OK. And is that a scheduled appointment?"

"I'm here to give a talk."

A light went on. "Oh, are you part of the drug lunch?"

Regardless of how I preferred to think of myself (an educator, a psychiatrist, a consultant), I was now classified as one facet of a lunch helping to pitch a drug, a convincing sidekick to help the sales rep. Eventually, with an internal wince, I began to introduce myself as "Dr. Carlat, here for the Wyeth lunch."

The drug rep who arranged the lunch was always there, usually an attractive, vivacious woman with platters of gourmet sandwiches in tow. Hungry doctors and their staff of nurses and receptionists would filter into the lunch room, grateful for free food.

Once there was a critical mass (and crucially, once the M.D.s arrived), I was given the go-ahead by the Wyeth reps to start. I dove into my talk, going through a handout that I created, based on the official slide deck. I discussed the importance of remission, the basics of the Thase study showing the advantage of Effexor, how to dose the drug, the side effects, and I added a quick review of the other common antidepressants.

While I still had some doubts, I continued to be impressed by the 10 percent advantage in remission rates that Effexor held over SSRIs; that advantage seemed significant enough to overcome Effexor's more prominent side effects. Yes, I was highlighting Effexor's selling points and playing down its disadvantages, and I knew it. But was my salesmanship going to bring harm to anybody? It seemed unlikely. The worst case was that Effexor was no more effective than anything else; it certainly was no less effective.

During my first few talks, I worried a lot about my performance. Was I too boring? Did the doctors see me as sleazy? Did the Wyeth reps find me sufficiently persuasive? But the day after my talks, I would get a call or an e-mail message from the rep saying that I did a great job, that the doctor was impressed and that they wanted to use me more. Indeed, I started receiving more and more invitations from other reps, and I soon had talks scheduled every week. I learned later that Wyeth and other companies have speaker-evaluation systems. After my talks, the reps would fill out a questionnaire rating my performance, which quickly became available to other Wyeth reps throughout the area.

As the reps became comfortable with me, they began to see me more as a sales colleague. I received faxes before talks preparing me for particular doctors. One note informed me that the physician we'd be visiting that day was a "decile 6 doctor and is not prescribing any Effexor XR, so please tailor accordingly. There is also one more doc in the practice that we are not familiar with." The term "decile 6" is drug-rep jargon for a doctor who prescribes a lot of medications. The higher the "decile" (in a range from 1 to 10), the higher the prescription volume, and the more potentially lucrative that doctor could be for the company.

A note from another rep reminded me of a scene from *Mission: Impossible*. "Dr. Carlat: Our main target, Dr., is an internist. He spreads his usage among three antidepressants, Celexa, Zoloft and Paxil, at about 25–30 percent each. He is currently using about 6 percent Effexor XR. Our access is very challenging with lunches six months out." This doctor's schedule of lunches was filled with reps from other companies; it would be vital to make our sales visit count.

Naïve as I was, I found myself astonished at the level of detail that drug companies were able to acquire about doctors' prescribing habits. I asked my reps about it; they told me that they received printouts tracking local doctors' prescriptions every week. The process is called "prescription data-mining," in which specialized pharmacy-information companies (like IMS Health and Verispan) buy prescription data from local pharmacies, repackage it, then sell it to pharmaceutical companies. This information is then passed on to the drug reps, who use it to tailor their drug-detailing strategies. This may include deciding which physicians to aim for, as my Wyeth reps did, but it can help sales in other ways. For example, Shahram Ahari, a former drug rep for Eli Lilly (the maker of Prozac) who is now a researcher at the University of California, San Francisco's School of Pharmacy, said in an article in the *Washington Post* that as a drug rep he would use this data to find out which doctors were prescribing

Prozac's competitors, like Effexor. Then he would play up specific features of Prozac that contrasted favorably with the other drug, like the ease with which patients can get off Prozac, as compared with the hard time they can have withdrawing from Effexor.

The American Medical Association is also a key player in prescription data-mining. Pharmacies typically will not release doctors' names to the data-mining companies, but they will release their Drug Enforcement Agency numbers. The AMA licenses its file of U.S. physicians, allowing the data-mining companies to match up DEA numbers to specific physicians. The AMA makes millions in information-leasing money.

Once drug companies have identified the doctors, they must woo them. In the April 2007 issue of the journal *PLoS Medicine*, Dr. Adriane Fugh-Berman of Georgetown teamed up with Ahari (the former drug rep) to describe the myriad techniques drug reps use to establish relationships with physicians, including inviting them to a speaker's meeting. These can serve to cement a positive relationship between the rep and the doctor. This relationship is crucial, they say, since "drug reps increase drug sales by influencing physicians, and they do so with finely titrated doses of friendship."

III. UNCOMFORTABLE MOMENTS

I gave many talks over the ensuing several months, and I gradually became more comfortable with the process. Each setting was somewhat different. Sometimes I spoke to a crowded conference room with several physicians, nurses, and other clinical staff. Other times, I sat at a small lunch table with only one other physician (plus the rep), having what amounted to a conversation about treating depression. My basic Effexor spiel was similar in the various settings, with the focus on remission and the Thase data.

Meanwhile, I was keeping up with new developments in the research literature related to Effexor, and not all of the news was positive. For example, as more data came out comparing Effexor with SSRIs other than Prozac, the Effexor remission advantage became slimmer—more like 5 percent instead of the originally reported 10 percent. Statistically, this 5 percent advantage meant that only one out of twenty patients would potentially do better on Effexor than SSRIs—much less compelling than the earlier proportion of one out of ten.

I also became aware of other critiques of the original Thase meta-analysis. For example, some patients enrolled in the original Effexor studies took SSRIs in the past and presumably had not responded well. This meant that the study population may have been enriched with patients who were treatment-resistant to SSRIs, giving Effexor an inherent advantage.

I didn't mention any of this in my talks, partly because none of it had been included in official company slides, and partly because I was concerned that the reps wouldn't invite me to give talks if I divulged any negative information. But I was beginning to struggle with the ethics of my silence.

One of my most uncomfortable moments came when I gave a presentation to a large group of psychiatrists. I was in the midst of wrapping up my talk with some information about Effexor and blood pressure. Referring to a large study paid for by Wyeth, I reported that patients are liable to develop hypertension only if they are taking Effexor at doses higher than three hundred milligrams per day.

"Really?" one psychiatrist in the room said. "I've seen hypertension at lower doses in my patients."

"I suppose it can happen, but it's rare at doses that are commonly used for depression."

He looked at me, frowned, and shook his head. "That hasn't been my experience."

I reached into my folder where I kept some of the key Effexor studies in case such questions arose.

According to this study of 3,744 patients, the rate of high blood pressure was 2.2 percent in the placebo group, and 2.9 percent in the group of patients who had taken daily doses of Effexor no larger than three hundred milligrams. Patients taking more than three hundred milligrams had a 9 percent risk of hypertension. As I went through the numbers with the doctor, however, I felt unsettled. I started talking faster, a sure sign of nervousness for me.

Driving home, I went back over the talk in my mind. I knew I had not lied—I had reported the data exactly as they were reported in the paper. But still, I had spun the results of the study in the most positive way possible, and I had not talked about the limitations of the data. I had not, for example, mentioned that if you focused specifically on patients taking between two hundred and three hundred milligrams per day, a commonly prescribed dosage range, you found a 3.7 percent incidence of hypertension. While this was not a statistically significant higher rate than the placebo, it still hinted that such moderate doses could, indeed, cause hypertension. Nor had I mentioned the fact that since the data were derived from placebo-controlled clinical trials, the patients were probably not representative of the patients seen in most real practices. Patients who are very old or who have significant medical problems are excluded from such studies. But real-world patients may well be at higher risk to develop hypertension on Effexor.

I realized that in my canned talks, I was blithely minimizing the hypertension risks, conveniently overlooking the fact that hypertension is a dangerous condition and not one to be trifled with. Why, I began to wonder, would anyone prescribe an antidepressant that could cause hypertension when there were many other alternatives? And why wasn't I asking this obvious question out loud during my talks?

I felt rattled. That psychiatrist's frown stayed with me—a mixture of skepticism and contempt. I wondered if he saw me for what I feared I had become—a drug rep with an M.D. I began to think that the money was affecting my critical judgment. I was willing to dance around the truth in order to make the drug reps happy. Receiving $750 checks for chatting with some doctors during a lunch break was such easy money that it left me giddy. Like an addiction, it was very hard to give up.

There was another problem: one of Effexor's side effects. Patients who stopped the medication were calling their doctors and reporting symptoms like severe dizziness and lightheadedness, bizarre electric-shock sensations in their heads, insomnia, sadness, and tearfulness. Some patients thought they were having strokes or nervous breakdowns and were showing up in emergency rooms. Gradually, however, it became clear that these were "withdrawal" symptoms. These were particularly common problems with Effexor because it has a short half-life, a measure of the time it takes the body to metabolize half of the total amount of a drug in the bloodstream. Paxil, another short half-life antidepressant, caused similar problems.

At the Wyeth meeting in New York, these withdrawal effects were mentioned in passing, though we were assured that Effexor withdrawal symptoms were uncommon and could usually be avoided by tapering down the dose very slowly. But in my practice, that strategy often did not work, and patients were having a very hard time coming off Effexor in order to start a trial of a different antidepressant.

I wrestled with how to handle this issue in my Effexor talks, since I believed it was a significant disadvantage of the drug. Psychiatrists frequently have to switch medications because of side effects or lack of effectiveness, and anticipating this potential need to change medications plays into our initial choice of a drug. Knowing that Effexor was hard to give up made me think twice about prescribing it in the first place.

During my talks, I found myself playing both sides of the issue, making sure to mention that withdrawal symptoms could be severe but assuring doctors that they could "usually" be avoided. Was I lying? Not really, since there were no solid published data, and indeed some patients had little problem coming off Effexor. But was I tweaking and pruning the truth in order to stay positive about the product? Definitely. And how did I rationalize this? I convinced myself that I had told "most" of the truth and that the potential negative consequences of this small truth "gap" were too trivial to worry about.

As the months went on, I developed more and more reservations about recommending that Effexor be used as a "first line" drug before trying the SSRIs. Not only were the newer comparative data less impressive, but the studies were short-term, lasting only six to twelve weeks. It seemed entirely possible that if the clinical trials had been longer—say, six months—SSRIs would have caught up with Effexor. Effexor was turning out to be an antidepressant that might have a very slight effectiveness advantage over SSRIs but that caused high blood pressure and had prolonged withdrawal symptoms.

At my next Lunch and Learn, I mentioned toward the end of my presentation that data in support of Effexor were mainly short-term, and that there was a possibility that SSRIs were just as effective. I felt reckless, but I left the office with a restored sense of integrity.

Several days later, I was visited by the same district manager who first offered me the speaking job. Pleasant as always, he said: "My reps told me that you weren't as enthusiastic about our product at your last talk. I told them that even Dr. Carlat can't hit a home run every time. Have you been sick?"

At that moment, I decided my career as an industry-sponsored speaker was over. The manager's message couldn't be clearer: I was being paid to enthusiastically endorse their drug. Once I stopped doing that, I was of little value to them, no matter how much "medical education" I provided.

IV. LIFE AFTER DRUG MONEY

A year after starting my educational talks for drug companies (I had also given two talks for Forest Pharmaceuticals, pushing the antidepressant Lexapro), I quit. I had made about $30,000 in supplemental income from these talks, a significant addition to the $140,000 or so I made from my private practice. Now I publish a medical-education newsletter for psychiatrists that is not financed by the pharmaceutical industry and that tries to critically assess drug research and marketing claims. I still see patients, and I still prescribe Effexor. I don't prescribe it as frequently as I used to, but I have seen many patients turn their lives around because they responded to this drug and to nothing else.

In 2002, the drug industry's trade group adopted voluntary guidelines limiting some of the more lavish benefits to doctors. While the guidelines still allow all-expenses-paid trips for physicians to attend meetings at fancy hotels, they no longer pay for spouses to attend the dinners or hand out tickets to musicals. In an e-mail message, a Wyeth spokesman wrote that Wyeth employees must follow that code and "our own Wyeth policies, which, in some cases, exceed" the trade group's code.

Looking back on the year I spent speaking for Wyeth, I've asked myself if my work as a company speaker led me to do bad things. Did I contribute to faulty medical decision making? Did my advice lead doctors to make inappropriate drug choices, and did their patients suffer needlessly?

Maybe. I'm sure I persuaded many physicians to prescribe Effexor, potentially contributing to blood-pressure problems and withdrawal symptoms. On the other hand, it's possible that some of those patients might have gained more relief from their depression and anxiety than they would have if they had been started on an SSRI. Not likely, but possible.

I still allow drug reps to visit my office and give me their pitches. While these visits are short on useful medical information, they do allow me to keep up with trends in drug marketing. Recently, a rep from Bristol-Myers Squibb came into my office and invited me to a dinner program on the antipsychotic Abilify.

"I think it will be a great program, Dr. Carlat," he said. "Would you like to come?" I glanced at the invitation. I recognized the name of the speaker, a prominent and widely published psychiatrist flown in from another state. The restaurant was one of the finest in town.

I was tempted. The wine, the great food, the proximity to a famous researcher—why not rejoin that inner circle of the select for an evening? But then I flashed to a memory of myself five years earlier, standing at a lectern and clearing my throat at the beginning of a drug-company presentation. I vividly remembered my sensations—the careful monitoring of what I would say, the calculations of how frank I should be.

"No," I said, as I handed the rep back the invitation. "I don't think I can make it. But thanks anyway."

Tina Rosenberg

When Is a Pain Doctor a Drug Pusher?

FROM THE *NEW YORK TIMES MAGAZINE*

Doctors who try to treat chronic pain are faced with a dilemma. If they underprescribe a pain medication out of fear of creating a dangerous dependency in the patient, treatment may not work. If they prescribe an amount that might seem high but proves effective, they can be liable to criminal prosecution, as Tina Rosenberg shows in the case of one such doctor, who is now in prison.

RONALD MCIVER IS A PRISONER in a medium-security federal compound in Butner, North Carolina. He is sixty-three years old, of medium height, and overweight, with a white Santa Claus beard, white hair, and a calm, direct, and intelligent manner. He is serving thirty years for drug trafficking, and so will likely live there the rest of his life. McIver (pronounced mi-KEE-ver) has not been convicted of drug trafficking in the classic sense. He is a doctor who for years treated patients suffering from chronic pain. At the Pain Therapy Center, his small storefront office not far from Main Street in Greenwood, South Carolina, he cracked backs, gave trigger-point injections, and put patients through physical therapy. He administered ultrasound and gravity-inversion therapy and devised exercise regimens. And he wrote prescriptions for high doses of opioid drugs like OxyContin.

McIver was a particularly aggressive pain doctor. Pain can be measured only by how patients say they feel: on a scale from zero to ten, a report of zero signifies the absence of pain; ten is unbearable pain. Many pain doctors will try to reduce a patient's pain to the level of five. McIver tried for a two. He prescribed more, and sooner, than most doctors.

Some of his patients sold their pills. Some abused them. One man, Larry Shealy, died with high doses of opioids that McIver had prescribed him in his bloodstream. In April 2005, McIver was convicted in federal court of one count of conspiracy to distribute controlled substances and eight counts of distribution. (He was also acquitted of six counts of distribution.) The jury also found that Shealy was killed by the drugs McIver prescribed. McIver is serving concurrent sentences of twenty years for distribution and thirty years for dispensing drugs that resulted in Shealy's death. His appeals to the U.S. Court of Appeals for the Fourth Circuit and the Supreme Court were rejected.

McIver's case is not simply the story of a narcotics conviction. It has enormous relevance to the lives of the one in five adult Americans who, according to a 2005 survey by Stanford University Medical Center, ABC News, and USA Today, reported they suffered from chronic pain—pain lasting for several months or longer. According to a 2003 study in the Journal of the American Medical Association, pain costs American workers more than $61 billion a year in lost productive time—and that doesn't include medical bills.

Contrary to the old saw, pain kills. A body in pain produces high levels of hormones that cause stress to the heart and lungs. Pain can cause blood pressure to spike, leading to heart attacks and strokes. Pain can also consume so much of the body's energy that the immune system degrades. Severe chronic pain sometimes leads to suicide. There are, of course, many ways to treat pain: some pain sufferers respond well to surgery, physical therapy, ultrasound, acupuncture, trigger-point injections, meditation, or over-the-counter painkillers like Advil (ibuprofen) or Tylenol (acetaminophen). But for many people in severe chronic pain, an opioid (an opiumlike compound) like Oxy-Contin, Dilaudid, Vicodin, Percocet, oxycodone, methadone, or morphine is the only thing that allows them to get out of bed. Yet most doctors prescribe opioids conservatively, and many patients and their families are just as cautious as their doctors. Men, especially, will simply tough it out, reasoning that pain is better than addiction.

It's a false choice. Virtually everyone who takes opioids will become physically dependent on them, which means that withdrawal symptoms like nausea and sweats can occur if usage ends abruptly. But tapering off gradually allows most people to avoid those symptoms, and physical dependence is not the same thing as addiction. Addiction—which is defined by cravings, loss of control, and a psychological compulsion to take a drug even when it is harmful—occurs in patients with a predisposition (biological or otherwise) to become addicted. At the very least, these include just below 10 percent of Americans, the number estimated by the United States Department of Health and Human Services to have active substance-abuse problems. Even a predisposition to addiction, however, doesn't mean a patient will become addicted to opioids. Vast numbers do not. Pain patients without prior abuse problems most likely run little risk. "Someone who has never abused alcohol or other drugs would be extremely unlikely to become addicted to opioid pain medicines, particularly if he or she is older," says Russell K. Portenoy, chairman of pain medicine and palliative care at Beth Israel Medical Center in New York and a leading authority on the treatment of pain.

The other popular misconception is that a high dose of opioids is always a dangerous dose. Even many doctors assume it; but they are nonetheless incorrect. It is true that high doses can cause respiratory failure in people who are not already taking the drugs. But that same high dose will not cause respiratory failure in someone whose drug levels have been increased gradually over time, a process called titration. For individuals who are properly titrated and monitored, there is no ceiling on opioid dosage. In this sense, high-dose prescription opioids can be safer than taking high doses of aspirin, Tylenol, or Advil, which cause organ damage in high doses, regardless of how those doses are administered. (Every year, an estimated five to six thousand Americans die from gastrointestinal bleeding associated with drugs like ibuprofen or aspirin, according to a paper published in the *American Journal of Gastroenterology.*)

Still, doctors who put patients on long-term high-dose opioids must be very careful. They must monitor the patients often to ensure that the drugs are being used correctly and that side effects like constipation and mental cloudiness are not too severe. Doctors should also not automatically assume that if small doses aren't working, that high doses will—opioids don't help everyone. And research indicates that in some cases, high doses of opioids can lose their effectiveness and that some patients are better off if they take drug "holidays" or alternate between different medicines. Pain doctors also concede that more studies are needed to determine the safety of long-term opioid use.

But with careful treatment, many patients whose opioid levels are increased gradually can function well on high doses for years. "Dose alone says nothing about proper medical practice," Portenoy says. "Very few patients require doses that exceed even two hundred milligrams of OxyContin on a daily basis. Having said this, pain specialists are very familiar with a subpopulation of patients who require higher doses to gain effect. I myself have several patients who take more than a thousand milligrams of OxyContin or its equivalent every day. One is a high-functioning executive who is pain-free most of the time, and the others have a level of pain control that allows a reasonable quality of life."

All modern pain-management textbooks advocate "titration to effect"—in other words, in cases where opioids are helping, gradually increasing the dosage until either the pain is acceptably controlled or the side effects begin to outweigh the pain-relief benefits. But the vast majority of doctors don't practice what the textbooks counsel. In part, this is because of the stigma associated with high-dose opioids, the fear that patients will become addicted and the fact that careful monitoring is very time-consuming. And most doctors have received virtually no training in medical school about managing pain: many hold the same misconceptions about addiction and dosage as the general public.

And even pain specialists can be conservative. Sean E. Greenwood died in August at age fifty of a cerebral hemorrhage that his wife, Siobhan Reynolds, attributes to untreated pain. Greenwood was seeing various pain specialists. What makes his undertreatment especially remarkable is that he and his wife founded the Pain Relief Network, an advocacy group that has been the most vocal opponent of prosecutions of doctors and financed part of the legal defense of many pain doctors. "Here I am—I know everyone, and even I couldn't get him care that didn't first regard him as a potential criminal," Reynolds said.

According to the pharmaceutical research company IMS Health, prescriptions for opioids have risen over the past few years. They are used now more than ever before. Yet study after study has concluded that pain is still radically undertreated. The Stanford University Medical Center survey found that only 50 percent of chronic-pain sufferers who had spoken to a doctor about their pain got sufficient relief. According to the American Pain Society, an advocacy group, fewer than half of cancer patients in pain get adequate pain relief.

Several states are now preparing new opioid-dosing guidelines that may inadvertently worsen undertreatment. This year, the state of Washington advised nonspecialist doctors that daily opioid doses should not exceed the equivalent of 120 milligrams of oral morphine daily—for oxycodone or OxyContin, that's just 80 milligrams per day—without the patient's also consulting a pain specialist. Along with the guidelines, officials published a statewide directory of such specialists. It contains twelve names. "There are just not enough pain specialists," says Scott M. Fishman, chief of pain medicine at the University of California, Davis, and a past president of the American Academy of Pain Medicine. And the guidelines may keep non-specialists from prescribing higher doses. "Many doctors will assume that if the state of Washington suggests this level of care, then it is unacceptable to proceed otherwise," Fishman says.

In addition to medical considerations real or imagined, there is another deterrent to opioid use: fear. According to the DEA, seventy-one doctors were arrested last year for crimes related to "diversion"—the leakage of prescription medicine into illegal drug markets. The DEA also opened 735 investigations of doctors, and an investigation alone can be enough to put a doctor out of business, as doctors can lose their licenses and practices and have their homes, offices, and cars seized even if no federal criminal charges are ever filed. Both figures—arrests and investigations—have risen steadily over the last few years.

Opioid drugs have been used to treat pain for decades, mostly for acute postsurgical pain or the pain of cancer patients. But in January 1996, Purdue Pharma helped increase the use of these drugs by introducing OxyContin—oxycodone with a time-release mechanism. Oncologists and pain doctors were the principal prescribers of opioids. But Purdue introduced the drug with an aggressive marketing campaign promoting OxyContin to general practitioners and the idea of opioid pain relief to doctors and consumers. The product's time-release mechanism, Purdue claimed, allowed steadier pain relief and deterred abuse.

Many pain sufferers found that OxyContin gave them better relief than they ever had before. But Purdue misrepresented the drug's potential for abuse. Last month, the company and three of its executives pleaded guilty to federal charges that they misled doctors and patients. The company agreed to pay $600 million in fines; and the executives, a total of $34.5 million. The pill's time-release mechanism turned out to be easily circumvented by crushing the pill and snorting or injecting the resulting powder.

By the late 1990s, OxyContin abuse was devastating small towns throughout Appalachia and rural New England. Pharmaceuticals, mainly opioids, are still widely abused—now more so than any illegal drug except marijuana. In 2005, according to the government's National Survey on Drug Use and Health, 6.4 million Americans, many of them teenagers, had abused pharmaceuticals recently. Most got the drug from friends or family—often, in the case of teenagers, from their parents' medicine cabinets.

At the time the OxyContin epidemic emerged, the DEA had far more experience seizing illegal drugs like cocaine and heroin. According to Mark Caverly, the head of the liaison and policy section for the DEA's Office of Diversion Control, the OxyContin epidemic, however, required the agency to step up its antidiversion efforts. In 2001 the DEA established the OxyContin Action Plan. The DEA dispatched investigators to the most troubled states and trained local law-enforcement officials.

THE BASIS OF THE PHYSICIAN-PATIENT relationship is trust. Trust is especially valued by pain patients, who often have long experience of being treated like criminals or hysterics. But when prescribing opioids, a physician's trust is easily abused. Pain doctors dispense drugs with a high street value that are attractive to addicts. All pain doctors encounter scammers; some doctors estimate that as many as 20 percent of their patients are selling their medicine or are addicted to opioids or other drugs. Experts are virtually unanimous in agreeing that even addicts who are suffering pain can be success-fully treated with opioids. Indeed, opioids can be lifesaving for addicts—witness the methadone maintenance therapy given to her-oin addicts. But treating addicts requires extra care.

Identifying the scammers is especially tricky because there is no objective test for pain—it doesn't show up on an X-ray. In one Brit-ish study, half the respondents who complained of lower-back pain had normal MRIs. Conversely, a third of those with no pain showed disk degeneration on their MRIs. The study suggested there could be a profound disconnection between what an MRI sees and what a patient feels.

There are red flags that indicate possible abuse or diversion: pa-tients who drive long distances to see the doctor, or ask for specific drugs by name, or claim to need more and more of them. But people with real pain also occasionally do these things. The doctor's di-lemma is how to stop the diverters without condemning other pa-tients to suffer unnecessarily, since a drug diverter and a legitimate patient can look very much alike. The dishonest prescriber and the honest one can also look alike. Society has a parallel dilemma: how to stop drug-dealing doctors without discouraging real ones and worsening America's undertreatment of pain.

IN JULY 2002, an insurance agent was sifting through records in Columbia, South Carolina, and paused at the file of one Larry Shealy. Shealy was getting OxyContin from a doctor named Ronald McIver—a lot of it. "The amounts were incredible; it jumped out in my face," the agent, who spoke on condition of anonymity, told me. "He was either selling them or taking so much he couldn't live." The agent did two things. He recommended to Shealy's employers that they exclude OxyContin coverage from their health insurance plan—which they did. And he called the DEA. Two days later, a DEA agent showed up in the insurance agent's office with an administrative subpoena to collect Shealy's file.

McIver wanted to be a doctor all his life, two of his daughters told me. But he taught and traveled for years before he finally enrolled at Michigan State University to become a D.O., or doctor of osteopathy, a more holistic alternative to a traditional medical education. (Osteopaths can do everything that traditional M.D.s can do, including prescribe opioids.) He began practicing pain medicine in the late 1980s. He had a practice in Florence, South Carolina, which ended when he declared bankruptcy in 2000. He moved to Greenwood to start over, establishing his new office in a storefront next to a chiropractor.

McIver was, by the account of his patients, an unusual doctor in the age of the ten-minute managed-care visit. He usually saw about six to twelve patients each day. One patient I spoke with—who never got high-dose opioids—said that his first visit with McIver lasted four hours, and in subsequent visits he spent an hour or more doing various therapies. Many patients said their visits lasted an hour. Patients taking opioids had to sign a pain contract and bring their pills in at each visit to be counted.

Many doctors take little interest in the administrative side of their practices, but McIver's neglect was epic. To save money, he employed mostly family. His wife, Carolyn, whose only medical training was from her husband, served as his assistant, giving shots and administering therapies. "His doctor's office did not resemble my family's doctor's office," said Sgt. Bobby Grogan, who was the investigator on the case for the Greenwood County sheriff. While McIver's treatment rooms were normal, his and his wife's offices—off limits to patients—were a mess, according to pictures presented at McIver's trial by Adam Roberson, the DEA's principal investigator. Used syringes, for example, overflowed their storage box. "His patient records were manila envelopes stuffed with receipts," Grogan told me.

When I interviewed him in prison recently, McIver told me that his records were complete but scattered. He said that he and his wife, distracted by a series of family tragedies, had employed a series of temporary receptionists who had botched the filing. He and his wife were trying to piece them together. "The records were probably half in the office and half at home for me to work on at night," he said. "I kept a box in the back of the car I worked on while Carolyn drove."

LESLIE SMITH FIRST CAME to see McIver in the fall of 2001. Smith was in his mid-forties and lived in Chapin, a small town near Columbia, a sixty-mile drive from Greenwood. He filled out a medical-history form and told McIver that his wrists hurt so badly that he was getting only three or four hours' sleep a night. He also said that a previous doctor helped him by prescribing OxyContin, and he mentioned the name of a doctor he said referred him. McIver examined Smith's wrists. Smith walked out with an opioid prescription and an appointment to come back the next week.

Smith's wrists did not hurt him, as he testified at McIver's trial. He was addicted to OxyContin and Dilaudid, which he injected. He complained of wrist pain because it was plausible: he had injured one wrist previously, requiring an operation that left scars, and he had arthritis in the other. Until June 2002, Smith kept getting prescriptions.

Smith saw McIver every few weeks. He testified that he had track marks on his arm at the time but always wore long sleeves to cover them. He said McIver never saw them. McIver put him on an electric nerve stimulator every visit for fifteen or thirty minutes on each hand and did osteopathic manipulations. He prescribed exercises. Smith bought a nerve-stimulator machine to use at home and told McIver it was helping. At McIver's request he filled out a pain chart and reported that his pain rated a five or six upon awakening, reached seven during the day and occasionally hit nine. "I answered all the questions exactly like I thought he'd want to hear them answered," Smith testified. At one point McIver found a syringe in Smith's pocket. Smith told McIver that he was going fishing later that day and that he used the syringe as part of his fishing equipment. That apparently satisfied McIver, who testified that his grandfather kept syringes in his tackle box to pump air into his bait.

Smith filled some of his prescriptions at the Hawthorne Pharmacy in West Columbia. There, Addison Livingston, the pharmacist, got suspicious. He noticed that Smith sometimes came in with other patients of McIver's, despite the fact that McIver worked nearly two hours' drive away. The patients obviously knew each other and would pick up large opioid prescriptions, paying cash and asking for brand-name drugs. Livingston called McIver, who confirmed he had written the prescriptions. At one point, McIver told Livingston that he, too, was suspicious, and that he had sent a letter about Smith to the state's Bureau of Drug Control.

In February 2002, McIver wrote to Larry McElrath, a BDC inspector, who read the letter at the trial. "Dear Larry," it read, "There are several people out of the Columbia/Chapin area who have aroused my curiosity about their use and possible misuse of medications. Some are referred by [another doctor] and seem legitimate. . . . They all pay cash despite some of them having insurance with prescription cards. . . . When they are in the office, they sometimes make a show of not knowing each other. . . . The situation is made complicated by the fact that each has some real pathology with objective findings that would justify the use of opiates if their pains are as bad as they say. I have given them the benefit of the doubt, but I'm becoming less inclined to do so. I would appreciate it if you could make some discrete inquiries and let me know whether my concerns are justified. . . . I certainly don't want to refuse help to someone who needs it. On the other hand, I want even less to be implicated in diversion or other improprieties." He listed their names and Social Security numbers.

McElrath did nothing with the letter. "It's incumbent upon the physician to have a trust with his patients," McElrath testified at the trial. "Here there was nothing that I could assume or conclude that any crimes had been committed."

Smith was the most damning of the several patients who testified against McIver. (Smith and the other patients mentioned here did not agree to be interviewed for this article, as they are suing McIver for alleged overprescription of addictive drugs. Such suits often prosper after successful criminal convictions, as civil suits are easier to win.) Smith had a confederate in Seth Boyer, who lived in Chapin and followed a similar pattern in his dealings with McIver: he exaggerated pains in his foot, never provided records from a previous doctor, and had needle tracks that he later testified McIver never saw. At one point, Boyer told McIver that he had spilled a bottle of liquid OxyFast, another opioid. (In reality, Boyer had injected it.) McIver wrote him a prescription for a replacement—apparently a violation of his standard pain-medication contract, which had a "no early refills" stipulation.

But McIver ended up discharging Boyer in June 2002, when Boyer altered a prescription so he could fill it three days early. He wrote McIver three pleading letters of protest, to no avail. "I was looking for an excuse to discharge them, and with Seth I found it," McIver told me. "I needed more than suspicion. With Les, he never actually did anything that allowed me to say, 'OK, here's that concrete piece of evidence.'"

McIver may have felt he needed more proof, but medically he probably had enough. Pain specialists told me that doctors can stop prescribing a drug whenever the risks outweigh the benefits, which includes the risk of abuse.

Another drug-dealing patient of McIver's was Kyle Barnes. She testified that she suffered from fibromyalgia, a chronic-pain syndrome, but exaggerated her pain to get higher levels of OxyContin and Roxicodone. She was addicted to those drugs before she began seeing McIver in July 2001. She also brought no medical records and drove three hours to each appointment. She got prescriptions on her second visit, during which McIver also did osteopathic manipulations and massage.

Barnes was in real pain. McIver did several different therapies at each visit. He set up an appointment for her at a sleep clinic, sent her for X-rays, and put a cast on her wrist. He knew she had trouble paying for her medicines, and he contacted Purdue Pharma to see if she qualified for reduced-price drugs. She kept claiming the drugs were not helping enough and was soon taking sixteen times the dose of OxyContin she took when she first saw him. One tip-off in her case should have been that she paid thousands of dollars a month in cash for her prescriptions, even though she was on Medicaid. She told McIver that her father and boyfriend were helping her buy them, which she later testified was partly true. But most of her income came from selling some of the drugs he prescribed, she testified.

In December 2003, McIver told her that he would stop treating her unless she took a drug screen. She did nothing. Three weeks later he told her again. She never returned.

Another patient whose story was particularly troubling was Barbee Brown. Brown was not a drug seeker but a genuine pain patient seeking relief from Reflex Sympathetic Dystrophy. McIver gave her very high doses of OxyContin right away, before she produced any records from other doctors. This was especially disturbing, because she had been addicted to crack cocaine for three months in the year before she came to him.

Brown saw McIver at least twice a week for six weeks. He did a thorough physical exam and took a complete history. He used many different kinds of therapies. But he also started her—someone who had never taken opioids—on forty-milligram pills of OxyContin and allowed her to control her own dosing schedule. "As long as you are not having side effects, do not be afraid to take the doses you need to get out of pain," he wrote to her. It was the same advice he gave many patients. "The number of milligrams does not matter. What matters is the number on the zero-to-ten scale."

The medicine helped. Brown testified that she ranked her pain at nine or ten when she first got to McIver. After seeing him, it dropped to a four. Her pain diary, which appears to be sincere, had various passages giving thanks that she met McIver. Brown did not become addicted. But allowing an opioid-naïve recovering crack addict to start on high-dose pills and control her own dosage, and telling her that her dosage didn't matter, seems reckless.

MCIVER'S THIRTY-YEAR SENTENCE was the result of the death of Larry Shealy, a fifty-six-year-old man who suffered intense back and knee pain, in addition to many other health problems. He first came to see McIver in February 2002, with full referrals and records. He was on OxyContin before seeing McIver but complained that his pain was still terrible, so McIver doubled his dose. This allowed Shealy to go back to work in an auto body shop.

Shealy was not a careful patient. A month after he started with McIver, he took fifteen OxyContin tablets in one day instead of the six he was prescribed. He was not harmed, but McIver testified that he asked Shealy to bring his family in so he could explain the dosing to them. At one point, McIver tried to taper down the OxyContin and replace it with methadone, but Shealy complained that the methadone made him drowsy. Shealy's son, David, an auto mechanic, testified that the OxyContin pain relief also came at a price. He said he felt his father was overmedicated—often sleepy. Once, his father backed his truck into a tree.

Shealy died in his sleep early on the morning of May 29, 2003. He had OxyContin pills in his stomach, and his bloodstream contained alprazolam—Xanax—as well. The pathologist at McIver's trial testified that the levels of drugs were consistent with the prescriptions McIver had been writing—the high levels that so alarmed the insurance agent. Shealy was taking five eighty-milligram tablets of OxyContin every twelve hours, plus up to six thirty-milligram tablets of Roxicodone every four hours for breakthrough pain, plus as much as two milligrams of alprazolam every eight hours. The prosecution's toxicologist, Demi Garvin, concluded that the OxyContin and Roxicodone caused Shealy's death by respiratory depression. The pathologist testified that she looked up this dosage and found it to be a fatal level.

But there is reason for doubt. According to Shealy's prescriptions, he had been taking the same dosage for at least two months, and possibly much longer. Pain specialists say that respiratory depression is extremely unlikely when dosage is consistent. In her testimony, Garvin agreed that what would be a toxic level in an opioid-naïve patient would be safe for someone titrated up properly. But she said she could not conclude he had been properly titrated, in part because she had not seen his medical records. Garvin declined to talk about the Shealy case with me because she is a witness for the Shealy family in their planned civil suit against McIver. But in a deposition for that lawsuit, she appeared to back away from blaming the Oxy-Contin. She described her view as: "Hey, there's a red flag here. This can certainly be your cause of death, but you need to go further in exploring whether or not it is."

There was something else that might have caused Shealy's death: he suffered from advanced congestive heart failure. The pathologist testified that he had 90 percent blockage in one coronary artery and 50 percent in another, and a greatly enlarged heart and other organs. He had a scar on the back wall of his heart that indicated he at one time suffered a heart attack. Opioids do not worsen heart disease and would likely have helped, because pain causes stress to the heart.

The testimonies of the patients Smith, Boyer, and Barnes were the parts of the trial that most directly addressed the question of whether McIver intentionally wrote prescriptions for a nonmedical purpose. This is the relevant legal test for the statute under which he was prosecuted. Several Supreme Court and district court cases have made it clear that under the Controlled Substances Act, a doctor is guilty of a crime if he intentionally acts as a drug pusher.

The judge in the McIver case, Henry F. Floyd, told the jurors that bad prescribing is the standard for malpractice, a civil matter. "That is not what we are talking about," he said. "We're not talking about this physician acting better or worse than other physicians." If McIver was a bad doctor—but still a doctor, with intent to treat patients—he was innocent. "If you find that a defendant acted in good faith in dispensing the drugs charged in this indictment, then you must find that defendant not guilty," Floyd said. But Floyd also told the jury to take bad doctoring into account in deciding McIver's intent.

This instruction—that bad doctoring does not prove intent but could be considered when weighing his intent—is subtle and potentially extremely confusing. It apparently confused the jurors. I spoke to two jurors, who told me their own views and characterized the jury discussion. The overwhelming factor, they said, was that McIver prescribed too much—the very red flag that alerted the insurance agent and set the case in motion.

The jurors I spoke with said that by far the most important testimony came from Steven Storick, a pain-management doctor in Columbia and the government's expert witness. Reviewing the records of patient after patient, Storick consistently testified that there were too many drugs. "This amount of medication is just extremely high in a situation like this," he said of one patient. This is "excessive," he said of another. "That's just an extremely high dose of drug," he said of a third. Storick, who declined to be interviewed for this article, testified that if he had a patient who exhibited no objective evidence of pain, he would not prescribe opioids. He would not have titrated patients as rapidly as McIver did or given them discretion. He disagreed with McIver's position that a doctor should try to bring a patient's chronic pain down to a level of two. He would stop titrating when a patient reached five out of ten.

The jurors took Storick's caution to heart, in part, they told me, because it resonated with their own experience with opioids and fears of addiction. I asked Jo Handy, a tall, elegant woman who is now thirty-nine and a real estate agent outside Greenville, why McIver was convicted. "It was the excessive prescriptions," she said in an interview in her office. "Excessive, and the number of them. I've been on some pain medication. But along with some other jurors we were, like, 'No—it's too much.'"

Handy said she knew McIver's treatment was excessive because

Storick said so, and because of her own experience. "Thirty counts is normal," she said. "He was giving sixty or ninety. A few of us had been on prescribed medicine. I had female issues. You as a person know not to take so much of that medication. If you were, you had a motive. Me, I still have a whole bottle left."

Christopher Poore, another juror, agreed that what swayed the jury was the volume of drugs prescribed. "The jury kept going back to the expert testimony of the prosecution's expert," he told me when I met him in Anderson, a town forty minutes from Greenwood. "It was beyond. It was too much." What should McIver have done, I asked, if he wanted to avoid jail? "He should have followed the convention more of what people are doing with pain medicine— not giving so much," Poore said.

Poore, who is forty and runs his family's heating and cooling business, described himself as the juror most skeptical of the prosecution's case. "There was another guy on the jury who said his sister-in-law had been taking pain pills and she had gotten addicted," Poore said. "He said I was taking up for McIver. I said, No, I'm taking up for you and me and anyone else who's on trial. I wanted to see rules, that this guy broke the rule. I never saw a rule he broke."

In the end Poore voted to convict. As is always the case, the jurors were dismissed before McIver was sentenced. Poore told me he supposed McIver was in prison. When I said McIver was serving thirty years, he looked shocked.

Interviews with jurors and the judge's sentencing decision indicated that photos of the messy conditions in McIver's and Carolyn's private offices also contributed to the impression that he was not a real doctor. Surprisingly, McIver's contacts with law enforcement— the letter about Smith and the others was one of several—helped the prosecution's case. "He called an officer about a patient," John P. Flannery II, McIver's appellate lawyer, explained to me. "There is no response. He gets zero. He took their silence as a sign everything was OK. They take that as knowledge of drug dealing." It mattered: the

Fourth Circuit's opinion rejecting McIver's appeal said, "That Appellant knew or suspected his patients of drug abuse is reflected by the fact that he wrote to state authorities to express concern that his patients might be selling their medication."

I asked Grogan, the local diversion investigator on the case, why he didn't follow up on McIver's suspicions. "I'm a cop, not a doctor," Grogan said. "I can't say to prescribe medication or not. How do I know he's not trying to fish me for information?"

"He doesn't have to call us to cut someone off," Mike Frederick, the chief deputy at the sheriff's office, told me. "This is no different than when regular illegal drug dealers will very often call us about other drug dealers. He did it most likely because he thought that person was a risk."

I had assumed that McIver's use of many different types of therapies would help his case, by showing he was not running a classic pill mill. But it may have hurt. During the appeal, the prosecutor William Lucius argued that the other treatments represented the profits of drug diversion. He addicted patients with high doses of opioids, Lucius contended, "so they would continue to come back to him" and "he could charge them for the treatments he gave."

HOW TYPICAL IS McIVER'S CASE? On the DEA's Web site the agency lists some of the doctors who have been prosecuted, and their crimes. There are some strikingly obvious and egregious cases of shady dealings: a doctor who wrote prescriptions in a gas station for a person who wasn't present; one who sold blank prescription forms; one who dispensed drugs to people who then shared them with him.

But not every doctor's intent to deal drugs is as clear. McIver was a crusader for high-dose opioids, credulous with patients and sloppy with documentation—a combination unwise in the extreme. But some of his patients said he was the only doctor who ever brought them relief. Prosecutors never brought any evidence that he intended to write prescriptions to be abused or sold. They never accused him of profiting from his patients' diversion except in collecting office fees. His patients who diverted or abused their opioids all testified they got their prescriptions by consistently lying to him. Nor is it convincing that his prescriptions killed Larry Shealy.

No one has analyzed the various prosecutions of pain doctors, so it is hard to determine how many of them look like McIver's. The DEA's list is incomplete. There have been many cases like McIver's, and most of these cases are not listed on the DEA's Web site. (One possible reason for this omission is that some of these cases are still being appealed.) And many cases that do appear on the list detail only vague crimes: convictions for prescribing "beyond the bounds of acceptable medical practice" or "dispensing controlled substances . . . with no legitimate medical purpose"—which is how the agency will most likely describe the McIver case if it ever includes the case on the list.

The DEA claims that it is not criminalizing bad medical decisions. For a prosecutable case, Caverly, the DEA officer, told me: "I need there to be no connection of the drug with a legitimate medical condition. I need the doctor to have prescribed the drug in exchange for an illegal drug, or sex, or just sold the prescription or wrote prescriptions for patients they have never seen, or made up a name."

I read this statement to Jennifer Bolen, a former federal prosecutor in drug-diversion cases who trained other prosecutors and now advises doctors on the law. "That's a good goal," she said. "I don't think they have yet reached that goal." McIver's case had no such broken connection, and in many cases the government has not produced testimony of intent to push drugs, providing evidence only of negligence or recklessness. In 2002, Bolen was one of the authors of a Justice Department document intended as part of a basic guide to prosecuting drug-diversion cases. The document, in the form of a reference card, dispenses with any need for a broken connection. It suggests that prosecutors need not prove a doctor had bad motives, that to be within the law a doctor had to prescribe "in strict compliance with generally accepted medical guidelines" and that doing an abbreviated medical history or physical examination is "probative" of lack of a legitimate medical purpose. The reference card was on the Justice Department's Web site but was pulled, according to the Pain Relief Network, which provided the card to me. Bolen told me: "I have no problem saying that if the card was all there was, it was not acceptable. But it isn't all there was." She described the card as one piece of a more thorough training, but added that many prosecutors followed its theories.

Prosecutors are in essence pressing jurors to decide whether an extra forty milligrams every four hours or a failure to X-ray is enough to send a doctor to prison for the rest of his life. One doctor, Frank Fisher, was arrested on charges that included the death of a patient taking opioids—who died as a passenger in a car accident. A Florida doctor, James Graves, is serving sixty-three years for charges including manslaughter after four patients overdosed on OxyContin he prescribed—all either crushed and injected their OxyContin or mixed it with alcohol or other drugs. "A lot of doctors are looking for safe harbor," Caverly said. "They want to know as long as they do A, B, C, D, or E, they're OK."

The DEA once thought that this was not an unreasonable desire. A few years ago, it worked with pain doctors to develop a set of frequently asked questions that set out what doctors needed to do to stay within the law. The FAQ recommended, for example, that doctors should do urine tests and discuss a patient's treatment with family and friends. In October 2004, the FAQ were erased from the agency's Web site. One reason was that one of their authors, who is a doctor, was about to use the list to testify on behalf of William Hurwitz, a pain doctor in McLean, Virginia. (Hurwitz was convicted on fifty counts of drug trafficking in 2004. His conviction was overturned, and he was recently retried and convicted on 16 lesser counts. He is awaiting sentencing.)

Caverly acknowledged the Hurwitz trial was one reason the FAQ were pulled, but said there were other reasons. He said such a check-off list could tie the DEA's hands. "Some doctor's going to pull that list of dos or don'ts out and say: 'See, I'm OK. I did these ten.' But there's a new wrinkle there—an eleventh one the doctor didn't do," he said. Most important, he went on to say, the FAQ had stepped over the line to insert the DEA into issues of medical practice. "We have to stay in our lane," he said. "Those definitions are the professional community's—not the DEA's."

In a perfect world, such reasoning would make sense. But the agency is defining issues of medical practice in dramatic fashion—by jailing doctors who step over the line. It would not seem to be bothering, however, to draw the line first.

The dilemma of preventing diversion without discouraging pain care is part of a larger problem: pain is discussed amid a swirl of ignorance and myth. Howard Heit, a pain and addiction specialist in Fairfax, Virginia, told me: "If we take the fact that 10 percent of the population has the disease of addiction, and if we say that pain is the most common presentation to a doctor's office, please tell me why the interface of pain and addiction is not part of the core curriculum of health care training in the United States?" Will Rowe, the executive director of the American Pain Foundation, notes that "pain education is still barely on the radar in most medical schools."

The public also needs education. Misconception reigns: that addiction is inevitable, that pain is harmless, that suffering has redemptive power, that pain medicine is for sissies, that sufferers are just faking. Many law-enforcement officers are as in the dark as the general public. Very few cities and only one state police force have officers who specialize in prescription-drug cases. Charles Cichon, executive director of the National Association of Drug Diversion Investigators (NADDI), says that NADDI offers just about the only training on prescription drugs and reaches only a small percentage of those who end up investigating diversion. I asked if, absent NADDI training, officers would understand such basics as whether there is a ceiling dose for opioids. "Probably not," he said.

There is another factor that might encourage overzealous prosecution: Local police can use these cases to finance further investigations. A doctor's possessions can be seized as drug profits, and as much as 80 percent can go back to the local police.

There are ways to prevent diversion without imprisoning doctors who have shown no illegal intent. They are increasingly used—but state authorities and doctors need to push even harder. The majority of states, South Carolina among them, do not yet have prescription monitoring—a central registry of prescriptions, which could help catch people getting opioids from several different doctors and pharmacies. Doctors should use more urine and blood tests, including screens that can tell quantities of drug present.

Last year, state medical boards took 473 actions against doctors for misdeeds involving prescribing controlled substances. In many cases, their licenses were pulled. Physicians can also lose their DEA registration, and with it the right to prescribe controlled substances. A few dozen do every year, although there is considerable overlap with medical-board actions. Washington is the first state to recommend that only pain specialists handle high-dose opioids; other states are likely to follow.

But such guidelines are futile while there is one pain specialist for, at the very least, every several thousand chronic-pain sufferers nationwide. And even though pain is an exciting new specialty, doctors are not flocking to it. The Federation of State Medical Boards calls "fear among physicians that they will be investigated, or even arrested, for prescribing controlled substances for pain" one of the two most important barriers to pain treatment, alongside lack of understanding. Various surveys of physicians have shown that this fear is widespread. "The bottom line is, doctors say they don't need this," said Heit. "They're in a health care system that wants them to see a patient every ten to fifteen minutes. They don't have time to take a complete history about whether the patient has been addicted. The fear is very real and palpable that if they prescribe Schedule II opioids they will come under the scrutiny of the DEA, and they don't need this aggravation."

Proper pain management will always take time, but the DEA can at least ensure that honest doctors need not fear prison. It should use the standard it claims to follow: for a criminal prosecution to occur, a doctor must have broken the link between the opioid and the medical condition. If the evidence is of recklessness alone, then it should be a case for a state medical board, the DEA's registration examiners, or a civil malpractice jury.

Undoubtedly, such a limit will allow a small group of pill-mill doctors to escape prison. But America lives with freeing suspects whose possible crimes are discovered through warrantless searches or torture—and unlike other suspects, doctors who lose their licenses are as incapacitated as those behind bars. For cases without the broken connection, prosecution is too blunt an instrument. It runs too high a risk of condemning innocent physicians to prison and discourages the practice of a medical specialty desperately needed by millions of Americans.

PAIN PATIENTS ARE THE COLLATERAL victims here. It is worth remembering that the vast majority of McIver's patients were not people who abused or sold their medicines. One of those who didn't was a man named Ben, a tall, heavy man in his fifties who lives about forty-five minutes from Greenwood. (He asked that his last name not be used because of the stigma still attached to taking opioid drugs.) Ben was once a mail carrier and a farmer and cattle rancher. But years of pushing eight-hundred-pound bales of hay wore out his back. In 2001 he had an operation to fuse the bottom three vertebrae. The few Vicodin his neurosurgeon prescribed did not control his pain. "I never had enough to get me through the night," he said. "He wasn't going to go any further than Vicodin—and he was doing me a favor by doing that, because his other partners wouldn't have done as much as he did." His neurosurgeon recommended he find a pain doctor. He started seeing McIver. The first examination, Ben said over coffee in a local Waffle House, was "extremely thorough—he had me crying. I hardly ever got out of there in less than two hours—he would be on top of me popping my back."

And he got opioids. With his typical imprudence, McIver told Ben: "You don't worry about it, take whatever you need to be pain-free, if it takes two pills or ten pills. If you're taking too much and slurring your words, you know to back off. Use some common sense." At McIver's request, Ben kept a diary of what he took and how much. He reached a top dosage of five eighty-milligram pills of OxyContin four times a day—more opioids than Shealy was taking at the time of his death. "I never felt high," he said. "They helped my pain. I could get out and work, use the bulldozer. I was working a 250-head cattle herd. I was doing everything relatively pain-free because of the drugs. They gave me my life back."

When McIver was closed down, Ben was lucky enough to have a family physician he knew well who took over his case. But the new doctor took a very different approach. Ben now gets three eighty-milligram pills of OxyContin a day, plus some breakthrough Roxicodone and eight hundred milligrams of Advil every four to six hours. "That's it and I'm very, very lucky to have it," he said. "My doctor is afraid they will say it's over the limit. I now get about three hours' sleep a night. I can stand for thirty minutes, maybe." He can no longer handle ranching and has sold his cattle. He considers himself retired.

With Ben's permission I talked to his current doctor, who said Ben was a good patient but had been taking way too much. "I thought Ben made an error," he said. "He had been taking five or six times the recommended dosage. There are well-recognized levels, and you don't step across the line. You may have to live with some pain."

Opioids have immense power—both to harm and to heal. They can be life-destroying, but high doses allowed Ben to work, to be with his family, to be who he is. In its prosecutions of pain doctors, the government fails to recognize the duality of these drugs. Ben's wife told me: "When Ben first went to Dr. McIver and filled out the form on what he used to be able to do and what he could do now, he cried. McIver said to him, 'I'm going to get you back to doing what you used to do.' And he did."

JEROME GROOPMAN

What's Normal?

FROM *THE NEW YORKER*

> *A child exhibits quicksilver changes in behavior, euphoric and gran-
> diose one minute, morose the next. A child merely acting like a kid?
> Or symptoms of bipolar disorder? Jerome Groopman investigates the
> controversy in diagnosing bipolar disorder in the very young.*

IN APRIL 2000, STEVEN HYMAN, a psychiatrist who at the
time was the director of the National Institute of Mental Health,
convened a meeting of nineteen prominent psychiatrists and
psychologists in order to discuss bipolar disorder in children. The
disorder has long been recognized as a serious psychiatric illness in
adults, characterized by recurring episodes of mania and depression.
(It is sometimes called manic depression.) People with bipolar disor-

der are often unable to hold down jobs; require lifelong treatment with powerful medications, many of which have severe side effects; and have high suicide rates. The disorder is thought to afflict between 1 and 4 percent of Americans and tends to run in families, although no genes for it have been identified. At the time of the meeting, few children had been given a diagnosis of the illness, and it was considered to begin, typically, in adolescence or early adulthood.

In the late 1990s, however, there was an increase in awareness of bipolar disorder in children, first in medical journals and then in places like BPParents, a Listserv founded by the mother of an eight-year-old boy who had been diagnosed with the disorder. Hyman himself had been consulted by parents of children who, he told me, were "really suffering and extremely disruptive, having violent outbursts at school and at home, and hard to contain under any circumstances." Many of the parents told Hyman that they believed their child had bipolar disorder, and they cited a book called *The Bipolar Child: The Definitive and Reassuring Guide to Childhood's Most Misunderstood Disorder.* The book, which was written by Demitri Papolos, a psychiatrist affiliated with the Albert Einstein College of Medicine, in New York City, and his wife, Janice, the author of several how-to manuals, had been published in 1999. (It has sold more than 200,000 copies, and a third edition came out last summer.) "The first parents who visited me came with the Papolos book in hand," Hyman said.

The Papoloses argued that bipolar disorder was often overlooked in children. In 1998, according to *The Bipolar Child*, nearly four million children were given Ritalin or other stimulants for hyperactivity; of that number, the Papoloses contended, more than a million would eventually receive a bipolar diagnosis. They also cited researchers' estimates that anywhere from a third to half of the 3.4 million children thought to suffer from depression were actually experiencing the early onset of bipolar disorder. The book detailed the negative effects of bipolar disorder on patients (disruptive behavior, drug abuse, suicide attempts) but also prominently featured what might be described as its paradoxical benefits:

> This illness is as old as humankind, and has probably been conserved in the human genome because it confers great energy and originality of thought. People who have had it have literally changed the course of human history: Manic-depression has afflicted (and probably fueled the brilliance of) people like Isaac Newton, Abraham Lincoln, Winston Churchill, Theodore Roosevelt, Johann Goethe, Honoré de Balzac, George Frederic Handel, Ludwig van Beethoven, Robert Schumann, Leo Tolstoy, Charles Dickens, Virginia Woolf, Ernest Hemingway, Robert Lowell, and Anne Sexton.

(These claims are similar to those made about other serious psychiatric disorders, particularly depression.) The Papoloses' research was based on responses to questionnaires that they distributed through BPParents, whose several hundred members are parents who suspect that their children have the disorder. "These children seem to burst into life and are on a different time schedule from the rest of the world right from the beginning," the Papoloses wrote. "Many are extremely precocious and bright—doing everything early and with gusto. They seem like magical children, their creativity can be astounding, and the parents speak about them with real respect, and sometimes even awe." The book included some parents' observations:

> She was always ahead of her time. She started talking at eight months with the words "kitty cat." She walked at nine months and was speaking in complete sentences by a year. She was writing small novels in the second grade. She acted and danced and sang way beyond her years.
>
> At eighteen months he climbed out of the baby bed in the middle of the night, opened the fridge, got out three dozen

eggs (it was Easter time), and proceeded to sit in his booster chair and crack three dozen eggs onto our hardwood floors. (He wanted to bake a real cake—he didn't like the toy mixing bowl I had given him to play with.) After the insurance company quit laughing they did pay to refinish our floors.

During the meeting at the NIMH, the psychiatrists and psychologists argued about whether bipolar disorder existed in children, and, if it did, how it could be distinguished from other syndromes affecting mood and behavior, such as attention deficit hyperactivity disorder (ADHD) and autism. One psychiatrist, Barbara Geller, a professor at Washington University in St. Louis, had published articles about children whose moods often fluctuated rapidly. In the course of a single day, the children were extremely sad, even suicidal, and then, suddenly, they became elated and "grandiose"—a term that psychiatrists use to mean an inflated sense of one's abilities. Geller believed that some of these children who matched several specific and narrow criteria had bipolar disorder. Joseph Biederman, a child psychiatrist at Massachusetts General Hospital, in Boston, who also attended the meeting, had treated children suffering from extreme symptoms of irritability and aggressive behavior and, often with a colleague, Janet Wozniak, had published several articles in medical journals asserting that these children met the criteria for bipolar disorder described in the *Diagnostic and Statistical Manual of Mental Disorders (DSM-IV)*, the reference book for psychiatric illnesses. Hyman encouraged the group to arrive at a consensus, in part to create a uniform set of criteria that could be used to enroll children in studies of the disorder.

In August 2001, the results of the meeting were published in the *Journal of the American Academy of Child and Adolescent Psychiatry*, and it was concluded that "bipolar disorder exists and can be diagnosed in prepubertal children," though the article went on to say that not all children who appeared to have the disorder satisfied the *DSM* criteria. The vagueness of the definition offered few guidelines for practical diagnosis.

Meanwhile, articles inspired by the Papoloses' book had begun appearing in newspapers and magazines, promoting the idea that there was a new diagnosis for troubled children. In August 2002, *Time* published a cover story titled "Young and Bipolar," with the tagline "Once Called Manic Depression, the Disorder Afflicted Adults. Now It's Striking Kids. Why?" The article featured a list of behaviors—adapted from the Papoloses' book—that was intended to help parents "recognize some warning signs" of the disorder. Among those were "poor handwriting," "complains of being bored," "is very intuitive or very creative," "excessively distressed when separated from family," "has difficulty arising in the A.M.," "elated or silly, giddy mood states," "curses viciously in anger," and "intolerant of delays." The magazine also published a sidebar listing prominent writers and musicians who may have suffered from bipolar disorder, including Lord Byron, Edgar Allan Poe, and Kurt Cobain. Although the article cited external factors such as stress and drug use, it also noted that the disorder is "hugely familial," as one doctor put it. (One mother, who was afflicted with bipolar disorder, claimed that she knew before her son was born that he would be bipolar, because he was restless even in the womb.)

Not long after the article came out, a research team at Massachusetts General Hospital, led by Biederman and Wozniak, began an eight-week comparative study of the antipsychotic drugs olanzipine (marketed under the name Zyprexa) and risperidone (Risperdal) for thirty-one children between the ages of four and six who had been given a diagnosis of bipolar disorder based on *DSM* criteria. During the trial, the children gained an average of six pounds and experienced sharp increases in prolactin, a pituitary hormone, which, when elevated, might interfere with sexual development. But their symptoms of severe irritability and aggression were markedly muted by the treatment, and the researchers, while noting the adverse effects, concluded that the drugs could be beneficial to bipolar children.

There are few reliable statistics on the incidence of pediatric bipolar disorder, but according to a national study of community-hospital discharge records, led by Brady Case, a research assistant professor of psychiatry at New York University, and Anthony Russo a child-psychiatry fellow at Bradley Hospital, in Providence, the percentage of mentally ill children under eighteen who have been given a diagnosis of the disorder increased more than fourfold between 1990 and 2000. Many doctors fear that the media, in drawing attention to bipolar disorder, may have exaggerated its prevalence in children and presented a misleading picture of the disorder. The situation has some similarities to the overdiagnosis of attention deficit disorder in the first half of the 1990s, during which the prescription of stimulants such as Ritalin tripled for children between the ages of two and four, according to a study published in February 2000, in the *Journal of the American Medical Association*. Some children do, of course, suffer from bipolar disorder, but it is important to recognize that the consequences of its treatment can be dire, particularly when parents are unaware of or ignore the dangerous side effects of the medications. In December 2006, a four-year-old girl in Massachusetts, who had received a bipolar diagnosis at the age of two and a half, died from an apparent overdose of Clonidine, a blood-pressure medicine used to sedate hyperactive children. She was also taking Seroquel, an antipsychotic, and Depakote, an antiseizure medication that helps regulate mood. (Her parents have been charged with murder and have pleaded not guilty.)

"The diagnosis has spread too broadly, so that powerful drugs are prescribed too widely," Hyman told me. "We are going to have hell to pay in terms of side effects."

ONE OF THE EARLIEST ACCOUNTS of bipolar disorder comes from Aretaeus the Cappadocian, a Greek physician who is believed to have practiced in Alexandria and Rome in the second century A.D. He wrote of the afflicted, "They are prone to change their mind readily; to become base, mean-spirited, illiberal, and in a little time, perhaps, simple, extravagant, munificent, not from any virtue of the soul, but from the changeableness of the disease. But if the illness becomes more urgent, hatred, avoidance of the haunts of men, vain lamentations; they complain of life, and desire to die." However, the disorder was not clearly recognized for centuries, and it wasn't until January 1854, at a meeting of the French Imperial Academy of Medicine, in Paris, that a physician named Jules Baillarger cited a mental illness that involved recurring oscillations between mania and depression: Baillarger described it as *folie à double forme* (dual-form insanity). The following month, another French doctor, Jean-Pierre Falret, described a similar illness to the academy, calling it *folie circulaire* (circular insanity). The term "manic-depressive psychosis" was introduced in 1896 by Emil Kraepelin, a German psychiatrist, who observed that periods of acute mania and depression were usually separated by longer intervals during which the patient was able to function normally.

Doctors made little progress in treating the disorder until after the Second World War, when John Cade, an Australian psychiatrist working at a veterans' hospital, set out to test the hypothesis that mania was related to a toxic buildup of urea in the bloodstream. By chance, he discovered that the lithium urate he injected into guinea pigs had a calming effect. After testing lithium carbonate on himself, he began administering it to his manic patients. It became the first successful drug therapy for a psychiatric disorder. (Lithium remained the only treatment for bipolar disorder for decades, and is still the most prevalent, but in recent years anticonvulsants and some antipsychotics have also proved effective.) In 1980, the term "bipolar disorder" replaced "manic-depressive disorder" as a diagnostic term in the *DSM*, but it was applied only to teenagers and adults.

"Until about ten years ago, it was considered quackery to talk about bipolar disorder in children," Barbara Geller told me. "The overwhelming number of adult and child psychiatrists believed that this was just a hyperactive child." Geller first encountered a child she believed exhibited the classic symptoms of bipolar disorder in the early '90s, a thirteen-year-old girl from a white middle-class family who was in the juvenile-correction system in the southern United States. The girl was euphoric despite her incarceration. "She seemed elated, grandiose, and infectiously funny, in spite of being in reform school," Geller recalled. Geller wondered whether the girl might be experiencing a manic episode, similar to those seen in adults with bipolar disorder. She began to interview other school-age and young adolescent children, seeking similar cases. One eleven-year-old girl harbored romantic fantasies about her teacher that led her to routinely disrupt class. She was also "delightfully euphoric" in an interview session with Geller, but as the questioning progressed she said that she had a loaded gun hidden at home, and had prepared a suicide note. Her parents searched their home, and found both the gun and the note. Geller was struck by the young girl's simultaneous grandiosity and depression; the two states are hallmarks of adult bipolar disorder, but they are rarely seen in such quick succession.

Geller found that the manner in which symptoms appeared in children with bipolar disorder was significantly different from that of most adults who had the illness. The episodes of mania and depression in most adults tend to subside after a few weeks or several months; children's episodes generally last longer, and cycle on a daily basis through a more extreme set of moods. "We have these kids who look so sad it hurts to watch them. And a moment later it looks like they've had a snort of coke," Geller said. "For four hours, they will be high: they are giggling, they are laughing, they are hypersexual, they want to touch the teacher, they want to undress in church, they talk too much, they sleep too little, and they think they are in charge of things. Then they switch. In the same day, they can suddenly become suicidal and depressed."

In 1995, with a grant from the NIMH, Geller began a longitudinal study of three groups of children: those she had diagnosed as having bipolar disorder using more precise categorical criteria than those specified in the *DSM*; those with attention deficit hyperactivity disorder; and a control group of children who had no known behavioral disorders. There were about ninety subjects in each group, and the average age was ten. Based on interviews with their parents and close relatives, Geller and her colleagues found that adult bipolar disorder was relatively common in the family members of the children who suffered from the disorder but not in those who had ADHD, or those in the control group. Geller concluded that there is a strong genetic basis for bipolar disorder in children, and that, among those diagnosed as having the disorder, more than 80 percent might also have ADHD.

Experts now agree that bipolar disorder can occur in children, but there is disagreement about which symptoms clearly indicate a diagnosis. Geller maintains that inappropriate euphoria and grandiose behavior must accompany symptoms of irritability or depression. Biederman and Wozniak contend that extreme irritability, including aggression, should compel a clinician to consider a diagnosis of pediatric bipolar disorder, in keeping with *DSM* criteria. However, Ellen Leibenluft, who heads the pediatric bipolar-disorder research program at the NIMH, told me that there is no certain way to classify even severe irritability as normal versus aberrant, particularly as children develop. Geller uses the analogy of sore throats: "Strep infection causes sore throat, but only 5 percent of all sore throats are due to strep, and 95 percent are due to viruses. Irritability is akin to the symptom of a sore throat: children with bipolar disorder are extremely irritable, but they comprise only a small subset of all irritable children."

Despite these differences, most researchers use the *DSM* criteria as a guideline. Demitri Papolos, however, argues against applying these categorical criteria, saying that their vagueness can cause confusion. "The diagnostic category in and of itself doesn't really capture the condition," he said. He prefers to make a diagnosis based on whether a patient's behavior matches the "core phenotype" he has developed, which includes mania and depression, among several other symptoms. "Once you see what this"—pediatric bipolar disorder—"looks like, you can't mistake it," he told me. "They call it the View. If you have the View, you get it. It's not apocalyptic, it's a very clear picture." Papolos, who is not a child psychiatrist, said that he has had children referred to him from all over the country, as many as two a week in the past seven years. He could not immediately recall any child in this group who did not have a bipolar diagnosis, because, he said, "the people who come to see me have read the book."

The need to establish diagnostic criteria is particularly urgent because many of the drugs given to bipolar children are relatively new and have not been tested extensively, especially in children. Depakote, the most common brand name for valproate, is an antiseizure medication for adults and children over the age of ten, which is also used to treat acute mania in adults; it can cause obesity and diabetes and has been associated with polycystic ovarian disease. The antipsychotic drug Risperdal can result in involuntary distorted movements, or "tardive dyskinesia." Lithium can cause decreased thyroid function and kidney failure. "Most important, we don't understand their long-term effects on the developing brain," Geller said. Failing to correctly diagnose pediatric bipolar disorder has its own dangers, since treating a bipolar patient with a selective serotonin reuptake inhibitor like Paxil or Zoloft, as if he were simply depressed, or with a stimulant like Ritalin, as if he had ADHD, might worsen his symptoms. Like other serious psychiatric illnesses, bipolar disorder is diagnosed largely by observing the patient's behavior. There is no blood test, or other clinical diagnostic tool, for the disorder; although brain scans have been performed on children who have been given the diagnosis, none have shown a definitive pattern.

Some books and articles on bipolar disorder in children and adolescents have suggested that a positive response to a drug like Risperdal, which can be effective in adults with manic bipolar disorder, indicates that the child is bipolar. In fact, the drugs typically given to bipolar children are what doctors call "nonspecific," which means that their apparent efficacy is not diagnostic of the syndrome. "All the medicines that work in bipolar cases also work in kids who are just aggressive," Geller said. "Children with mental retardation who acted aggressively were treated with drugs like lithium, and it helped to mute their behavior. But it also made them very thirsty, so they started drinking from toilet bowls and engaging in other kinds of unsuitable behavior. The contention that treatment with these drugs 'makes' the diagnosis is frightening—and completely untrue."

In January 2007, the American Academy of Child and Adolescent Psychiatry published a paper to guide clinicians in their assessment and treatment of children and adolescents with bipolar disorder. The paper cited a survey of members of the Illinois-based Child and Adolescent Bipolar Foundation, in which 24 percent of the children from 854 families who had been given a diagnosis of bipolar disorder were between the ages of zero and eight. (A more recent survey conducted by the foundation puts the number at 15 percent.) "The validity of diagnosing bipolar disorder in preschool children has not been established," the academy's paper noted. "Until the validity of the diagnosis is established in preschoolers, caution should be taken before making the diagnosis in anyone younger than age six. The evidence is not yet sufficient to conclude that most presentations of juvenile mania are continuous with the classic adult disorder." Biederman and Wozniak have given the diagnosis to preschool children and have included them in drug trials. But other experts, Geller and Leibenluft among them, contend that bipolar disorder cannot yet be accurately diagnosed in a child younger than six, because there is currently no consensus on what constitutes aberrant behavior at that age. In addition, they say, symptoms of manic behavior must be elicited through an interview not only with the parents but also with the children themselves; those younger than six may lack the language to describe what they are experiencing.

IN THE EARLY '90, in an effort to insure that children were receiving the correct diagnosis, Geller established a second-opinion clinic for bipolar disorder at Washington University. "Following the publication of the Papoloses' book, we began to have a greater influx of people into the clinic," she said. The positive effect of the book, she added, was that "parents realized it was OK to take their kids to a child psychiatrist." At the same time, the book could lead to false diagnoses. Geller went on, "In the clinic, the first question we have learned to ask of parents is 'Have you read the Papoloses' book?' And 'What in the book resembles your child?' And we will get answers like 'My child is irritable and he likes sweets.'" Geller's team developed stringent criteria to characterize mania as abnormal elation and grandiosity—such as inappropriate bouts of extreme giddiness, or hyperbolic statements of one's importance or ability—so that irritability alone was not adequate to establish a diagnosis of bipolar disorder. Many parents, she said, cling to a bipolar diagnosis when, in fact, the child is suffering from an autistic developmental disorder: "Wouldn't you rather have your child grow up to be Ted Turner," who has bipolar disorder, "than Rain Man?"

April Prewitt, a child psychologist who trained at Harvard and practices in Lexington, Massachusetts, also spends a good deal of time "undiagnosing" children who have been told they are bipolar. In the past three years, Prewitt says, she has seen thirty children and adolescents diagnosed as having bipolar disorder. In her opinion, only two had the malady. "It has become a diagnosis *du jour*, as ADHD was five years ago," Prewitt told me. "Not only is the diagnosis being made incorrectly but it's being made in younger and younger children." She said that parents routinely arrive at her office with the Papolos book, and with lists of behaviors like the one featured in *Time*. "Each one of these could be behaviors due to something completely different," she said. "I could score twenty on this list on a bad day."

Prewitt recalled a seven-and-a-half-year-old boy she saw, who lived in an affluent Boston suburb. Max (a pseudonym) had trouble concentrating and was refusing to go to school. His pediatrician had diagnosed bipolar disorder and begun treating him with Risperdal and Seroquel. "It turned out that the diagnosis was 'a divorce situation,'" Prewitt said. Max's parents had separated and were undergoing bitter divorce negotiations. "Max had put on twenty pounds because of the medication, while he was being shuttled back and forth, one week with mom and one week with dad." Prewitt believed that the parents' feuding was causing Max to oscillate between being sullen and withdrawn and aggressive and hyperactive. She recommended that Max be evaluated by a neuropsychologist, who found that he had only some minor attention deficits. During the following six months, his parents went into mediation in an effort to settle their divorce more amicably, and Max was weaned off his medications.

Prewitt maintains that it may not be possible to diagnose bipolar disorder with certainty in a preadolescent child. "After all these years, I am not sure of the diagnosis of bipolar disorder until a child is well into adolescence," she told me. "I've never seen a seven- or eight-year-old that I would be comfortable definitively diagnosing with bipolar disorder. The changes that children undergo, both in the biology of their development and in the need to adapt to changes in environment at home and at school—interactions with parents, siblings, and other children—all can trigger behaviors with rapid and wild swings of mood."

PHILLIP BLUMBERG, A PSYCHOTHERAPIST in Manhattan, told me, "Psychological diagnosis is, in essence, a story. If you have a mood disorder, there is the fear, the shame, and the confusion—the stigma—associated with it, so you want to grab on to the most concrete and clear story you can. There is something about the clarity of bipolar disease, particularly its biological basis, which is incredibly soothing and seductive."

Blumberg, who for two years was a vice president at ABC Motion Pictures, believes that advertising by pharmaceutical companies has influenced the public's view of bipolar disorder. (Eli Lilly, in particular, has come under fire for its marketing practices. The drug company is currently the subject of lawsuits that claim that the company attempted to hide Zyprexa's side effects, and promoted the drug for off-label uses. Lilly has denied the accusations.) Blumberg described recent ads, for drugs like Zyprexa, that include a list of symptoms characteristic of the disorder. "But, of course, we all have these symptoms," he said. "Sometimes we're irritable. Sometimes we're excited and elated, and we don't know why. With every form of advertising, the first goal is to make people feel insecure. Usually, they are made to feel insecure about their smell or their looks. Now we are beginning to see this in psychiatric advertising. The advertisements make frenetic, driven parents feel insecure about the behavior of their children."

Blumberg noted that he had seen instances of the disorder in some children, and that it was a real and serious diagnosis. But he also cited the mounting pressure on children, particularly in the middle and upper classes, to succeed, first at private or selective public schools, and then at exclusive colleges and universities. "These kids become very well turned-out products," he said. "They live to have résumés. They don't have résumés because they live." Parents may fear that children who behave in an eccentric way are at a disadvantage, and in turn pressure the pediatrician or the psychiatrist to come up with a diagnosis and offer a treatment. "Then an industry grows up around it. This, then, enters as truth in the popular imagination."

The debate over pediatric bipolar disorder will likely extend to the next edition of the *Diagnostic and Statistical Manual of Mental Disorders*. "*DSM* always has an out in its definitions, a category called NOS—'not otherwise specified,' " Steven Hyman said. "The problem with describing a kid who is up-and-down and irritable and sullen and wild and then grandiose is that he could indeed be rapidly cycling between mania and depression, but it could be an awful lot of other things, too. Bipolar disorder in children represents the intersection of two great extremes of ignorance: how to best treat bipolar disorder and how to treat children for anything. It's really important that we define the kids with bipolar disorder and treat them, but it's also important that we not begin to diagnose kids with excess exuberance or moodiness as having the disease. We have to realize that we are risking treating children who could turn into obese diabetics with involuntary movements. There is something very real about the kids with devastating and disruptive symptoms, but the question is still the boundaries. You can do more harm than good if you treat the wrong kid."

SALLY SATEL

Supply, Demand, and Kidney Transplants

FROM *POLICY REVIEW*

> *When the psychiatrist Sally Satel was told by her doctors that she would need a kidney transplant, she experienced first-hand the dilemmas and anxieties caused by a demand for donor organs that far outstrips supply. Fortunately for Satel, her good friend Virginia Postrel donated one of her kidneys; but, as Satel reasons, altruism is not enough to solve the organ-shortage problem.*

I N MAY 2002, CLOIS GUTHRIE, an eighty-five-year-old retired osteopathic surgeon, got the phone call he was waiting for: A suitable kidney had just become available for him. A renal transplant would mean liberation from the dialysis machine to which he had been tethered for two miserable years. Elated,

Guthrie and his wife raced to the Porter Adventist Medical Center from their home in north Denver one hundred miles away.[1] Yet mere hours before the operation was to take place, the center's transplant surgeons were engaged in anguished deliberation over whether Guthrie was actually the right person to get that kidney. He was, after all, eighty-five years old. How much longer would he live with a new kidney? Shouldn't the organ, taken from a healthy thirty-year-old motorcyclist who had died from head trauma, be given to a younger person who would get many more years of life from it?

The doctors decided to proceed, but in the end, there was a technical glitch and the operation did not take place. Guthrie went back on dialysis. Two-and-a-half years later he was dead of a heart attack at age eighty-eight. The ethical dilemma sparked by his case, however, did not die with him. Indeed, the question of "how old is too old for a transplant?" is being asked with increasing urgency by transplant professionals as the chasm between supply and demand widens inexorably.

Uneasy questions of allocation arise in environments of scarcity. Who will get to stay on the crowded lifeboat and who will be tossed overboard? This age-old tension between utility to society—the maximum good for the maximum number—and fairness to the individual is notoriously hard to resolve. In the case of the shortage of transplantable kidneys, it is made gratuitously more difficult by a "transplant community" that resists experimenting with bold ideas to increase the supply.

DIRE SHORTAGE, MODEST REMEDIES

Organ transplantation is one of the crowning achievements of medical science. Yet from 1954—the year of the first renal transplant—to

---

[1] Dr. Guthrie's story was reported in Alan Zarembo, "How Old Is Too Old for a Transplant? Kidneys Are Scarce. Elderly Patients May Get Fewer if Rules Change," *Los Angeles Times* (November 5, 2006).

the present, there have never been enough organs to meet demand. The dearth includes all transplantable organs—hearts, livers, lungs, pancreases—but because dialysis can keep patients with renal failure alive, the shortage of kidneys is most acute in terms of volume. Indeed, over three-quarters of the national wait-list population comprises those waiting for a kidney.

In January 2007, roughly 70,000 people were waiting for a kidney, according to the United Network for Organ Sharing (UNOS), which maintains the national registry of transplant candidates under monopoly contract with the Department of Health and Human Services. In big cities, where the ratio of needy patients to available organs is highest, the wait for a kidney ranges from five to eight years. This time is spent undergoing dialysis, a procedure that circulates the patient's blood through a machine (once called an "artificial kidney") that purifies it, siphons off accumulated water, and returns it to the body. Patients typically visit a dialysis center for treatment three times a week, for four hours each time. Many patients are deeply ambivalent about dialysis. They acknowledge its life-preserving role yet resent it as a vast intrusion into daily life that is often uncomfortable and debilitating.

Commitment to dialysis ends when a patient receives a transplant or dies. In 2006, only 17,804 people—or about one-quarter of the population waiting at the beginning of the year—received kidneys. Meanwhile, over 3,813 died waiting and 1,190 became too sick to transplant. It is a grim picture that is guaranteed to worsen. By 2010, the median waiting time is projected to be at least ten years long, extending well beyond the length of time that most adults, especially those over sixty-five, are able to survive on dialysis.

Despite decades of public education about the virtues of donating organs at death, the supply of cadaver organs has remained disappointingly steady over the years. Half of all Americans have designated themselves as donors on their driver's licenses or on state-run donor registries, yet if family members are unaware of a loved one's preference, they are just as likely as not to grant permission for the organs to be taken. This is understandable considering that the request generally comes at the worst possible time. Not only is their relative dead, but he may well have met a sudden and violent end. It is just this kind of victim—a young, healthy individual with severe head trauma (think helmetless motorcycle accident)—who is the optimal deceased donor.

In response to the crying demand, transplant centers have been rethinking the definition of a transplantable organ. Consequently, kidneys from so-called expanded-criteria donors have become an important new source. These are deceased donors, over sixty years of age, or those between fifty and fifty-nine years who had hypertension in life or died of a stroke. The retrieval of organs from moribund patients who may never meet the criteria for brain death is also being advanced. Such potential donors are called "non-heart-beating" because they died of cardiac arrest. Finally, there is an effort to boost the numbers of transplants using kidneys from living donors through so-called kidney exchanges. In such an arrangement, two or more prospective donor-recipient couples who are incompatible with each other match with a member of the other pair.

Zero-Sum Game

Much as these procurement innovations are welcome, they are likely to make only a modest contribution to supply.[2] The burden on transplant bureaucrats at UNOS, then, is how to allocate the limited reserve of cadaver kidneys in order to make optimal use of

[2] Paired exchanges are estimated to yield about a thousand new kidneys per year, according to Dorry Segev, M.D., Johns Hopkins School of Medecine (personal communication, May 16, 2007). Non-heart-beating decedents may yield as many as seven thousand new donors per year (at 1.5 kidneys per donor, plus other vital organs) from deaths that occur in hospitals, according to Jim Warren, editor of *Transplant News* (personal communication, April 12, 2007).

them. It is a problem UNOS has been working on for some time, and last winter the agency sponsored its Public Forum to Discuss Kidney Allocation Policy to unveil its new proposal. Central to the new allocation model was the concept of "Life Years from Transplant"—that is, how much longer a patient would likely survive after a transplant compared to how long he would live if he continued on dialysis instead.

The agency was seeking to maximize the number of additional years lived. In comparison with the first come, first served rule that currently guides kidney allocation, a scheme that prioritizes on the basis of life-years gained is more utilitarian. Starkly put, it views a healthy organ as wasted if it outlives its recipient—and the goal of optimizing longevity is to avoid a Clois Guthrie situation, in which a young kidney, able to prolong by decades the life of a forty-year-old, instead goes to an eighty-five-year-old who dies a few years later, taking the organ with him.

In addition to penalizing older candidates, justifiable though this may be, the new UNOS proposal also disfavors candidates who have waited the longest. Indeed, it is the very fact of having waited years that weakens a candidate's prospect for receiving a kidney. This is because the more time he spends on dialysis, the more medical deterioration he suffers and, as a result, the fewer years of added life a new kidney will confer.

Proponents of the new UNOS scheme denied that a rigid age cut-off would be applied. Yet they acknowledged that older candidates, as a class, would indeed be disadvantaged. This was a flashpoint. A number of physicians and transplant recipients took the microphone during the comment period to object. A few predicted the American Association of Retired Persons would lobby strongly against the proposal, as the new plan appeared to "discriminate" against older people. Others speculated that the thriving international black market in kidneys would get a further boost from well-to-do elderly who would become "transplant tourists"—the term used to describe patients who go overseas to purchase a transplant on the black market.

One transplant recipient in the audience expressed worry that patients who had logged years on the list already would become demoralized if waiting time became a minor factor in assignment. "It will destroy hope," she said; knowing you will get a kidney "is what keeps you going." Another recipient wondered: "Who's to say an older person's five years of life are any less important than a younger person's nine years? . . . That's playing God and people aren't going to like it."[3]

That is not playing God; that is playing man—the all-too-human affair of people deliberating strenuously and in good faith to determine what is right. The UNOS meeting dramatized the timeless "tragic choice" dilemma. The phrase comes from a classic 1978 book called *Tragic Choices* (W. W. Norton & Company) by esteemed legal theorists Guido Calabresi and Philip Bobbitt. The authors delineate the conflicts society faces when it is compelled to distribute limited resources. Most wrenching for citizens and policymakers are choices among fundamental values. A transplant surgeon who cares for patients in a milieu of scarcity is no less a healer. He still wants his patients—young and old—to receive kidneys. Without question, a transplant will afford them a better quality and quantity of life—irrespective of age—not only in terms of liberation from dialysis but because they will be spared its cardiovascular complications. Yet, on the other hand, the surgeon is torn between his duty to the patient before him and the utilitarian imperative of enhancing survival benefit across the population of patients needing transplants.

It is the eternal trade-off that comes with medical rationing: individual versus societal benefit. Who will be saved? This question re-

<hr>

[3] Judith Graham, "Should Age Determine Who Gets a Kidney Transplant? Controversial Proposal Would Put Younger Patients Higher on Waiting List," *Chicago Tribune* (February 9, 2007).

calls the classic quandary of lifeboat ethics that famously confronted the physicians who developed chronic dialysis in the early 1960s.

RATIONING AND ITS DISCONTENTS

Effective dialysis began during World War II, but the technology could be used only in patients with temporary damage to their kidneys. Ongoing dialysis was not possible because of the difficulty of maintaining a connection between patients' veins and arteries and the artificial kidney, as the dialysis machine was once called. Glass tubing was used as the conduit for blood as it flowed into and out of the machine, but it would get blocked with clots from the vessel in which it was embedded. Doctors rotated the tubes after each dialysis session, but within a few weeks all viable vasculature would clot off. Unless the patient's renal failure reversed before then, he would die.

A breakthrough came on March 9, 1960, when a nephrologist at the University of Washington experimented with tubes made of a relatively new substance called Teflon. The nonstick surface of the Teflon allowed the tubes to remain in the patient's arm for months without clotting. "Suddenly, we took something that was 100 percent fatal and overnight turned it into 90 percent survival," said Dr. Belding Scribner, who pioneered the technique. And just as suddenly, the University of Washington Hospital was inundated with referrals for dialysis from physicians and patients across the country. Scribner secured private funding to underwrite an experimental dialysis program, at the Seattle Artificial Kidney Center. It opened in January 1962 with a total of three treatment slots. But who among the dying should get them? Scribner argued that the job of choosing among medically eligible candidates ought to be shared by society.

Thus, a lay committee, known as the Admissions and Policy Committee of the Seattle Artificial Kidney Center at Swedish Hospital, was formed to decide nothing less than who would be allowed to live. The committee comprised seven volunteers—a lawyer, minister, housewife, state government official, labor leader, banker, and surgeon—and was among the earliest instances, if not the first, of physicians bringing nonprofessionals into the realm of clinical decision-making. Committee members insisted on remaining anonymous so that the medical staff, the public, and especially the applicant-patients would never know their identities.

The lay committee took its Solomonic charge seriously. "As human beings ourselves," the lawyer told a reporter, "we rejected the idea, instinctively, of classifying other human beings in pigeonholes, but we realized we had to narrow the field somehow." The committee did this by considering many factors, among them the applicant's income, sex, marital status, net worth, nature of occupation, extent of education, church attendance, number of dependents (more dependents conferred a better chance of being chosen), and potential for rehabilitation.

Within five months of its founding, the lay committee was thrust into the public eye. The *New York Times* ran a front page story in May 1962: "Panel Holds Life-or-Death Vote in Allocating Artificial Kidney." In November, *Life*, the 1960s' most influential popular weekly magazine, ran a story by journalist Shana Alexander called "They Decide Who Lives, Who Dies: Medical Miracle Puts Moral Burden on a Small Community." The exposé drew national attention to what was happening in Seattle. Alexander dubbed it the "Life or Death Committee," and the accompanying photo spread depicted the members in silhouette, as if sitting in harsh judgment. In 1965, Edwin Newman narrated an NBC documentary about Seattle called "Who Shall Live?" Vocal physicians, social scientists, theologians, and legal scholars felt that the selection of dialysis recipients based upon determinations of human worth was an affront to the ideal of equality. The moral claim of each patient to treatment was equivalent, they argued. One much-cited essay in the *UCLA Law Review* bitingly observed that "The Pacific Northwest is no place for a Henry David Thoreau with bad kidneys," chiding the committee for ruling out creative nonconformists.

Allocation of dialysis was among the first tragic choices to arise in the modern era of medical innovation. Only on the battlefield had case-by-case triage ever been performed so explicitly. The scarcity of dialysis treatment generated layer upon layer of vexing questions. Who should receive life-saving care, who should choose, and on what principle? Some medical centers employed a first come, first served policy; a few shuttered their programs altogether so they would not have to choose. Most medical centers favored the utilitarian principle of maximizing outcome—in other words, choosing patients who would get more productive years out of dialysis. Inevitably, this ended up favoring the same patients who impressed selection committees as more conscientious, better educated, and more likely to be beneficiaries of the emotional and instrumental support that comes along with stable families.

Thus, throughout the 1960s and early '70s, the tragic choice posed by dialysis was uneasily resolved in favor of handpicking patients. By 1972, however, pressure from advocates and physicians had become strong enough to move Congress to establish universal funding for dialysis. The Medicare End-Stage Renal Disease (ESRD) program, the new federal entitlement, offered virtually unfettered access to dialysis. Once the gate had swung open— admitting not only more patients who were good dialysis candidates (i.e., otherwise fairly healthy and cooperative) but also those with other medical and behavioral problems—enrollment and costs skyrocketed. More and more patients were kept alive on dialysis, and as advances in antirejection medication came about in the early 1980s, more people wanted kidney transplants, viewing dialysis as a bridge to surgery. With each passing year, the volume of candidates burgeoned and the amount of time spent waiting for a cadaver kidney increased. These dynamics were soon compounded by the aging of the U.S. population and the surge of diabetes (the most common single cause of renal failure). In 2004, the most recent year for which there are data, the ESRD enrolled about 340,000 patients and cost over eighteen billion dollars. Less than 1 percent of the

Medicare population consumed more than 6 percent of Medicare expenditures that year.

Today, thirty-five years after the establishment of ESRD, we have come full circle. Once we were rationing dialysis, and now we are rationing kidneys. The details have changed, but the basic challenge of assigning scarce resources has not.

A Shortage of Altruism

Under the 1984 National Organ Transplant Act, anyone who offers or receives something of material value in exchange for an organ can be charged with a felony. The ban's rationale was twofold: to prevent lurid scenarios in which desperately poor people auctioned off their spare parts to the wealthy and to ensure that citizens had equal access to the organs collected. "The prisoner in California gets the heart transplant because he needs it and is first on the list. It's blind to whether you're a saint or a sinner or a celebrity. That's key to maintaining the public trust," said Mark Fox, former head of the UNOS ethics committee.

But the trust is already damaged because of the death toll over which UNOS presides. The equity that UNOS seeks to preserve is "degenerating into an equal opportunity to die waiting," nephrologist Benjamin Hippen told the President's Council on Bioethics last year. The dire shortage of organs today is striking evidence of the fact that altruism is not sufficient to produce enough organs. In 2006, there were 7,180 deceased donors (yielding an average of 1.5 kidneys each) and 6,242 living donors (mainly family and friends). At the end of that year, the 67,000 candidates remaining dwarfed the number of available organs.

A cohort of physicians and economists has sought for at least two decades to persuade the transplant establishment to apply incentives to increase the organ supply. Many creative arrangements—from tax credits to tuition vouchers for children to charitable contributions in the donor's name—should be given a trial, they urge, to see whether new practices could compensate for the limits of altruism. Over the past few years, their voices have grown more insistent. In 2003, the American Medical Association testified in favor of a House bill that proposed pilot studies of incentives for harvesting the organs of deceased donors. At the 2006 World Transplant Congress in Boston, Dr. Richard Fine, president of the American Society of Transplantation, asked his colleagues: "Is it wrong for an individual who wishes to utilize part of his body for the benefit of another [to] be provided with financial compensation that could obliterate a life of destitution for the individual and his family?"

At the 2007 annual meeting of the American Society of Transplant Surgeons in January, a straw poll revealed that 80 to 85 percent of participants were in favor of studying incentives for living and deceased donors, according to society president Dr. Arthur Matas of the University of Minnesota. The public is receptive as well. A 2007 national Gallup Poll on attitudes toward donation of organs after death found that incentives would encourage more respondents to donate than would be discouraged from doing so, though the majority said their decision would remain unchanged. Most striking, among the 18- to 34-year-old age group, 34 percent said they would be more likely to donate and 6 percent said they would be less likely to do so; the rest were unchanged.

THE MORAL IMPERATIVE TO INNOVATE

Paradoxically, the current system, which is based on an altruism-or-else policy, undermines respect for individual autonomy, one of the most dearly held values in bioethics. Why shouldn't donors be able to receive some form of reward for giving up a kidney to save a life, especially if the act of compensating them encourages others to do the same? To be sure, the idea of combining organ donation with material gain can make people queasy. Yet the mix of financial and humanitarian motives is commonplace. No one objects, for example, to a tax credit for charitable contributions—a financial incentive to complement the "pure" motive of giving to others. The great teachers who enlighten us and the doctors who heal us inspire no less gratitude because they are paid. A salaried firefighter who saves a child trapped in a burning building is no less heroic in our eyes. Motives for giving an organ should not be the issue. More important than whether a kidney is given freely or for material gain is that it will increase the supply of kidneys to ameliorate suffering.

How could incentives work? A plan first offered in the late 1980s proposed to register a would-be donor today in return for the possibility of a much larger payment to his estate should his organ be used at his death. A major advantage of such a forward-looking approach is that the decision-making burden is taken off family members at a painful time—when they are sitting in the emergency room learning that someone they love is now brain-dead. The drawback to this plan, however, is that deceased donors alone cannot meet the need for kidneys because very few Americans who die, perhaps thirteen thousand a year (or less than 1 percent of all deaths), possess organs healthy enough for transplanting. Thus, even if every American participated in a futures arrangement, the need for thousands of kidneys would go unmet.

Of course, the organs from deceased donors obtained through a futures mechanism would be a most welcome contribution to the pool, but in order to enhance supply more quickly and more robustly, living donors need to be recruited as well. Not only can a person live a healthy life with one kidney, but the long-term risk to the donor is negligible, and short-term risks are comparable to surgical intervention in general (about three deaths in ten thousand). Moreover, a kidney from a living donor lasts longer than one from a cadaver, thereby keeping young and middle-aged recipients from getting back in the queue for a second organ sooner than they otherwise might have to.

What kinds of incentives could be offered to individuals amenable to relinquishing a kidney while living? Perhaps the federal government could offer lifetime Medicare coverage. Not only would this be medically responsible; it could serve as an inducement to donate—not to mention saving taxpayer dollars by liberating patients from costly dialysis. If the promise of health insurance did not attract a sufficient pool of donors, other incentives could be offered as well. For example, the donor could choose from a menu of options including a deposit to a 401(k) retirement plan, tax credits, tuition vouchers for the donor's children, long-term nursing care, family health coverage, life and nonfatal injury insurance, a charitable contribution in the donor's name, or cash payments stretched over time. Under this scheme, Medicare would underwrite the incentives in light of the fact that it already pays for dialysis treatment, which costs about $66,650 per patient annually according to the United States Renal Data Service. Compare these expenses with the cost of a transplant operation—approximately $75,000 in all for the one-time cost of the surgeries and hospital stays of the donor and recipient, plus the first year of follow-up medical care, including medication. Thereafter, the cost of immunosuppressive drugs is about $12,000 per year.[4]

An important concern accompanying any enrichment plan is the potential for exploiting donors—especially low-income donors who, as the critics reasonably claim, will be the most likely to find incentives attractive. This is why donor protection is the linchpin of any compensation model. Standard guidelines for physical and psychological screening, donor education, and informed consent could be formulated by a medical organization, such as the American Society of Transplant Surgeons, or another entity designated by the federal Department of Health and Human Services (HHS). A "waiting period" of three to six months could be built in to ensure that the prospective donor has ample time to think through his commitment. Monitoring donor health post-transplant is important as well and

---

[4] Arthur J. Matas and Mark Schnitzler, "Payment for Living Donor (Vendor) Kidneys: A Cost-Effectiveness Analysis," *American Journal of Transplantation* 4 (February 2004).

should include annual physicals and laboratory tests for one to two years after donation.

Incentive arrangements could be overseen by HHS or an entity it designates. As is currently the case with cadaver organs, kidneys obtained from compensated donors would be matched with the next best candidate waiting on the national list. This would require revising the National Organ Transplant Act to lift the ban on valuable consideration so that experimental trials could be conducted. Alternatively, Congress might permit individual states to apply for a waiver from the ban in order to devise their own incentive systems.

Within such a framework, altruistic donation would proceed in parallel with a system that offers compensation. Any medical center or physician that objects to the practice of compensating donors could simply opt out of performing transplants that use such organs. Recipients on the list are free to turn down a paid-for organ and wait for one given altruistically. Choice for all—donors, recipients, and physicians—is enhanced, while lives are saved.

A REPRIEVE

These broad proposals and variants on them need considerable elaboration. There is no denying the political and practical challenges that come with introducing payment into a twenty-year-old scheme built on the premise that generosity is the only legitimate motive for relinquishing an organ. Yet as death and suffering mount, constructing an incentive program to increase the supply of transplantable organs becomes a moral imperative.

We see again in 2007 how remarkable advances in transplant medicine have reintroduced the classic dilemma of equity versus utilitarianism that beset the medical profession in 1962 after the advent of dialysis therapy. To be sure, the world of health care is no stranger to rationing—simply determining what pharmaceuticals are to be covered by health plans is a form of rationing. But rarely has case-by-case access to treatment been as overt as it was in the early days of dialysis and is now in transplantation, though these days it is perhaps somewhat less dramatic, as choices are not based explicitly upon patients' social characteristics.

Would incentives work? There is good reason to be optimistic, but pilot studies are required to test various models. Architects of any new plan must give serious consideration to principled reservations and practical concerns; but they must act nonetheless, taking small, cautious steps. One thing is certain: A larger pool of kidneys would offer a reprieve, at least a partial one, to those patients languishing on dialysis and to those given the tragic charge of deciding which lives will be saved.

OLIVER SACKS

The Abyss

FROM *THE NEW YORKER*

Oliver Sacks meets a man, a once brilliant musician, who, due to an illness, lost his ability to process his experiences into memories; for him, everything happens as if for the first time. And yet he retains much of his musical proficiency.

IN MARCH OF 1985, CLIVE WEARING, an eminent English musician and musicologist in his mid-forties, was struck by a brain infection—a herpes encephalitis—affecting especially the parts of his brain concerned with memory. He was left with a memory span of only seconds—the most devastating case of amnesia ever recorded. New events and experiences were effaced almost instantly. As his wife, Deborah, wrote in her 2005 memoir, *Forever Today:*

His ability to perceive what he saw and heard was unimpaired. But he did not seem to be able to retain any impression of anything for more than a blink. Indeed, if he did blink, his eyelids parted to reveal a new scene. The view before the blink was utterly forgotten. Each blink, each glance away and back, brought him an entirely new view. I tried to imagine how it was for him. . . . Something akin to a film with bad continuity, the glass half empty, then full, the cigarette suddenly longer, the actor's hair now tousled, now smooth. But this was real life, a room changing in ways that were physically impossible.

In addition to this inability to preserve new memories, Clive had a retrograde amnesia, a deletion of virtually his entire past.

When he was filmed in 1986 for Jonathan Miller's extraordinary documentary *Prisoner of Consciousness*, Clive showed a desperate aloneness, fear, and bewilderment. He was acutely, continually, agonizingly conscious that something bizarre, something awful, was the matter. His constantly repeated complaint, however, was not of a faulty memory but of being deprived, in some uncanny and terrible way, of all experience, deprived of consciousness and life itself. As Deborah wrote:

> It was as if every waking moment was the first waking moment. Clive was under the constant impression that he had just emerged from unconsciousness because he had no evidence in his own mind of ever being awake before. . . . "I haven't heard anything, seen anything, touched anything, smelled anything," he would say. "It's like being dead."

Desperate to hold on to something, to gain some purchase, Clive started to keep a journal, first on scraps of paper, then in a notebook. But his journal entries consisted, essentially, of the statements "I am awake" or "I am conscious," entered again and again every few minutes. He would write: "2:10 P.M: This time properly awake. . . .

2:14 P.M: this time finally awake. . . . 2:35 P.M: this time completely awake," along with negations of these statements: "At 9:40 P.M. I awoke for the first time, despite my previous claims." This in turn was crossed out, followed by "I was fully conscious at 10:35 P.M., and awake for the first time in many, many weeks." This in turn was canceled out by the next entry.

This dreadful journal, almost void of any other content but these passionate assertions and denials, intending to affirm existence and continuity but forever contradicting them, was filled anew each day, and soon mounted to hundreds of almost identical pages. It was a terrifying and poignant testament to Clive's mental state, his lostness, in the years that followed his amnesia—a state that Deborah, in Miller's film, called "a never-ending agony."

Another profoundly amnesic patient I knew some years ago dealt with his abysses of amnesia by fluent confabulations. He was wholly immersed in his quick-fire inventions and had no insight into what was happening; so far as he was concerned, there was nothing the matter. He would confidently identify or misidentify me as a friend of his, a customer in his delicatessen, a kosher butcher, another doctor—as a dozen different people in the course of a few minutes. This sort of confabulation was not one of conscious fabrication. It was, rather, a strategy, a desperate attempt—unconscious and almost automatic—to provide a sort of continuity, a narrative continuity, when memory, and thus experience, was being snatched away every instant.

Though one cannot have direct knowledge of one's own amnesia, there may be ways to infer it: from the expressions on people's faces when one has repeated something half a dozen times; when one looks down at one's coffee cup and finds that it is empty; when one looks at one's diary and sees entries in one's own handwriting. Lacking memory, lacking direct experiential knowledge, amnesiacs have to make hypotheses and inferences, and they usually make plausible ones. They can infer that they have been doing *something*, been *somewhere*, even though they cannot recollect what or where. Yet Clive, rather than making plausible guesses, always came to the conclusion that he had just been "awakened," that he had been "dead." This seemed to me a reflection of the almost instantaneous effacement of perception for Clive—thought itself was almost impossible within this tiny window of time. Indeed, Clive once said to Deborah, "I am completely incapable of thinking."

At the beginning of his illness, Clive would sometimes be confounded at the bizarre things he experienced. Deborah wrote of how, coming in one day, she saw him

> holding something in the palm of one hand, and repeatedly covering and uncovering it with the other hand as if he were a magician practising a disappearing trick. He was holding a chocolate. He could feel the chocolate unmoving in his left palm, and yet every time he lifted his hand he told me it revealed a brand new chocolate.
>
> "Look!" he said. "It's new!" He couldn't take his eyes off it.
>
> "It's the same chocolate," I said gently.
>
> "No . . . look! It's changed. It wasn't like that before . . ." He covered and uncovered the chocolate every couple of seconds, lifting and looking.
>
> "Look! It's different again! How do they do it?"

Within months, Clive's confusion gave way to the agony, the desperation, that is so clear in Miller's film. This, in turn, was succeeded by a deep depression, as it came to him—if only in sudden, intense, and immediately forgotten moments—that his former life was over, that he was incorrigibly disabled.

As the months passed without any real improvement, the hope of significant recovery became fainter and fainter, and toward the end of 1985 Clive was moved to a room in a chronic psychiatric unit—a room he was to occupy for the next six and a half years but which he was never able to recognize as his own. A young psychologist saw Clive for a period of time in 1990 and kept a verbatim record of everything he said, and this caught the grim mood that had taken hold. Clive said at one point, "Can you imagine one night five years long? No dreaming, no waking, no touch, no taste, no smell, no sight, no sound, no hearing, nothing at all. It's like being dead. I came to the conclusion that I was dead."

The only times of feeling alive were when Deborah visited him. But the moment she left, he was desperate once again, and by the time she got home, ten or fifteen minutes later, she would find repeated messages from him on her answering machine: "Please come and see me, darling—it's been ages since I've seen you. Please fly here at the speed of light."

To imagine the future was no more possible for Clive than to remember the past—both were engulfed by the onslaught of amnesia. Yet, at some level, Clive could not be unaware of the sort of place he was in, and the likelihood that he would spend the rest of his life, his endless night, in such a place.

But then, seven years after his illness, after huge efforts by Deborah, Clive was moved to a small country residence for the brain-injured, much more congenial than a hospital. Here he was one of only a handful of patients, and in constant contact with a dedicated staff who treated him as an individual and respected his intelligence and talents. He was taken off most of his heavy tranquillizers, and seemed to enjoy his walks around the village and gardens near the home, the spaciousness, the fresh food.

For the first eight or nine years in this new home, Deborah told me, "Clive was calmer and sometimes jolly, a bit more content, but often with angry outbursts still, unpredictable, withdrawn, spending most of his time in his room alone." But gradually, in the past six or seven years, Clive has become more sociable, more talkative. Conversation (though of a "scripted" sort) has come to fill what had been empty, solitary, and desperate days.

THOUGH I HAD CORRESPONDED with Deborah since Clive first became ill, twenty years went by before I met Clive in person. He was so changed from the haunted, agonized man I had seen in Miller's 1986 film that I was scarcely prepared for the dapper, bubbling figure who opened the door when Deborah and I went to visit him in the summer of 2005. He had been reminded of our visit just before we arrived, and he flung his arms around Deborah the moment she entered.

Deborah introduced me: "This is Dr. Sacks." And Clive immediately said, "You doctors work twenty-four hours a day, don't you? You're always in demand." We went up to his room, which contained an electric organ console and a piano piled high with music. Some of the scores, I noted, were transcriptions of Orlandus Lassus, the Renaissance composer whose works Clive had edited. I saw Clive's journal by the washstand—he has now filled up scores of volumes, and the current one is always kept in this exact location. Next to it was an etymological dictionary with dozens of reference slips of different colors stuck between the pages and a large, handsome volume, *The 100 Most Beautiful Cathedrals in the World*. A Canaletto print hung on the wall, and I asked Clive if he had ever been to Venice. No, he said. (Deborah told me they had visited several times before his illness.) Looking at the print, Clive pointed out the dome of a church: "Look at it," he said. "See how it soars—like an angel!"

When I asked Deborah whether Clive knew about her memoir, she told me that she had shown it to him twice before, but that he had instantly forgotten. I had my own heavily annotated copy with me, and asked Deborah to show it to him again.

"You've written a book!" he cried, astonished. "Well done! Congratulations!" He peered at the cover. "All by you? Good heavens!" Excited, he jumped for joy. Deborah showed him the dedication page: "For my Clive." "Dedicated to me?" He hugged her. This scene was repeated several times within a few minutes, with almost exactly the same astonishment, the same expressions of delight and joy each time.

Clive and Deborah are still very much in love with each other, despite his amnesia. (Indeed, Deborah's book is subtitled *A Memoir of Love and Amnesia.*) He greeted her several times as if she had just arrived. It must be an extraordinary situation, I thought, both maddening and flattering, to be seen always as new, as a gift, a blessing.

Clive had, in the meantime, addressed me as "Your Highness" and inquired at intervals, "Been at Buckingham Palace? . . . Are you the prime minister? . . . Are you from the UN?" He laughed when I answered, "Just the U.S." This joking or jesting was of a somewhat waggish, stereotyped nature and highly repetitive. Clive had no idea who I was, little idea who anyone was, but this bonhomie allowed him to make contact, to keep a conversation going. I suspected he had some damage to his frontal lobes, too—such jokiness (neurologists speak of *Witzelsucht,* joking disease), like his impulsiveness and chattiness, could go with a weakening of the usual social frontal-lobe inhibitions.

He was excited at the notion of going out for lunch—lunch with Deborah. "Isn't she a wonderful woman?" he kept asking me. "Doesn't she have marvellous kisses?" I said yes, I was sure she had.

As we drove to the restaurant, Clive, with great speed and fluency, invented words for the letters on the license plates of passing cars: "JCK" was Japanese Clever Kid; "NKR" was New King of Russia; and "BDH" (Deborah's car) was British Daft Hospital, then Blessed Dutch Hospital. *Forever Today,* Deborah's book, immediately became "Three-Ever Today," "Two-Ever Today," "One-Ever Today." This incontinent punning and rhyming and clanging was virtually instantaneous, occurring with a speed no normal person could match. It resembled Tourettic or savantlike speed, the speed of the preconscious, undelayed by reflection.

When we arrived at the restaurant, Clive did all the license plates in the parking lot and then, elaborately, with a bow and a flourish, let Deborah enter: "Ladies first!" He looked at me with some uncertainty as I followed them to the table: "Are you joining us, too?"

When I offered him the wine list, he looked it over and exclaimed, "Good God! Australian wine! New Zealand wine! The colonies are producing something original—how exciting!" This partly indicated his retrograde amnesia—he is still in the 1960s (if he is anywhere), when Australian and New Zealand wines were almost unheard of in England. "The colonies," however, was part of his compulsive waggery and parody.

At lunch he talked about Cambridge—he had been at Clare College, but had often gone next door to King's, for its famous choir. He spoke of how after Cambridge, in 1968, he joined the London Sinfonietta, where they played modern music, though he was already attracted to the Renaissance and Lassus. He was the chorus master there, and he reminisced about how the singers could not talk during coffee breaks; they had to save their voices ("It was often misunderstood by the instrumentalists, seemed standoffish to them"). These all sounded like genuine memories. But they could equally have reflected his knowing *about* these events, rather than actual memories of them—expressions of "semantic" memory rather than "event" or "episodic" memory. Then he spoke of the Second World War (he was born in 1938) and how his family would go to bomb shelters and play chess or cards there. He said that he remembered the doodlebugs: "There were more bombs in Birmingham than in London." Was it possible that these were genuine memories? He would have been only six or seven, at most. Or was he confabulating or simply, as we all do, repeating stories he had been told as a child?

At one point, he talked about pollution and how dirty petrol engines were. When I told him I had a hybrid with an electric motor as well as a combustion engine, he was astounded, as if something he had read about as a theoretical possibility had, far sooner than he had imagined, become a reality.

In her remarkable book, so tender yet so tough-minded and realistic, Deborah wrote about the change that had so struck me: that Clive was now "garrulous and outgoing . . . could talk the hind legs off a donkey." There were certain themes he tended to stick to, she said, favorite subjects (electricity, the Tube, stars and planets, Queen Victoria, words and etymologies), which would all be brought up again and again:

"Have they found life on Mars yet?"

"No, darling, but they think there might have been water . . ."

"Really? Isn't it amazing that the sun goes on burning? Where does it get all that fuel? It doesn't get any smaller. And it doesn't move. We move round the sun. How can it keep on burning for millions of years? And the Earth stays the same temperature. It's so finely balanced."

"They say it's getting warmer now, love. They call it global warming."

"No! Why's that?"

"Because of the pollution. We've been emitting gases into the atmosphere. And puncturing the ozone layer."

"OH NO! That could be disastrous!"

"People are already getting more cancers."

"Oh, aren't people stupid! Do you know the average IQ is only 100? That's terribly low, isn't it? One hundred. It's no wonder the world's in such a mess."

Clive's scripts were repeated with great frequency, sometimes three or four times in one phone call. He stuck to subjects he felt he knew something about, where he would be on safe ground, even if here and there something apocryphal crept in. . . . These small areas of repartee acted as stepping stones on which he could move through the present. They enabled him to engage with others.

I would put it even more strongly and use a phrase that Deborah used in another connection, when she wrote of Clive being poised upon "a tiny platform . . . above the abyss." Clive's loquacity, his almost compulsive need to talk and keep conversations going, served to maintain a precarious platform, and when he came to a stop the abyss was there, waiting to engulf him. This, indeed, is what happened when we went to a supermarket and he and I got separated briefly from Deborah. He suddenly exclaimed, "I'm conscious now. . . . Never saw a human being before . . . for thirty years. . . . It's like death!" He looked very angry and distressed. Deborah said the staff calls these grim monologues his "deads"—they make a note of how many he has in a day or a week and gauge his state of mind by their number.

Deborah thinks that repetition has slightly dulled the very real pain that goes with this agonized but stereotyped complaint, but when he says such things she will distract him immediately. Once she has done this, there seems to be no lingering mood—an advantage of his amnesia. And, indeed, once we returned to the car Clive was off on his license plates again.

BACK IN HIS ROOM, I spotted the two volumes of Bach's *Fortyeight Preludes and Fugues* on top of the piano and asked Clive if he would play one of them. He said that he had never played any of them before, but then he began to play Prelude 9 in E Major and said, "I remember this one." He remembers almost nothing unless he is actually doing it; then it may come to him. He inserted a tiny, charming improvisation at one point, and did a sort of Chico Marx ending, with a huge downward scale. With his great musicality and his playfulness, he can easily improvise, joke, play with any piece of music.

His eye fell on the book about cathedrals, and he talked about cathedral bells—did I know how many combinations there could be with eight bells? "Eight by seven by six by five by four by three by two by one," he rattled off. "Factorial eight." And then, without pause: "That's forty thousand." (I worked it out, laboriously: it is 40,320.)

I asked him about prime ministers. Tony Blair? Never heard of him. John Major? No. Margaret Thatcher? Vaguely familiar. Harold Macmillan, Harold Wilson: ditto. (But earlier in the day he had seen a car with "JMV" plates and instantly said, "John Major Vehicle"—showing that he had an *implicit* memory of Major's name.) Deborah wrote of how he could not remember *her* name, "but one day someone asked him to say his full name, and he said, 'Clive

David Deborah Wearing—funny name that. I don't know why my parents called me that.'" He has gained other implicit memories, too, slowly picking up new knowledge, like the layout of his residence. He can go alone now to the bathroom, the dining room, the kitchen—but if he stops and thinks en route he is lost. Though he could not describe his residence, Deborah tells me that he unclasps his seat belt as they draw near and offers to get out and open the gate. Later, when he makes her coffee, he knows where the cups, the milk, and the sugar are kept. He cannot *say* where they are, but he can go to them; he has actions, but few facts, at his disposal.

I decided to widen the testing and asked Clive to tell me the names of all the composers he knew. He said, "Handel, Bach, Beethoven, Berg, Mozart, Lassus." That was it. Deborah told me that at first, when asked this question, he would omit Lassus, his favorite composer. This seemed appalling for someone who had been not only a musician but an encyclopedic musicologist. Perhaps it reflected the shortness of his attention span and recent immediate memory—perhaps he thought that he had in fact given us dozens of names. So I asked him other questions on a variety of topics that he would have been knowledgeable about in his earlier days. Again, there was a paucity of information in his replies and sometimes something close to a blank. I started to feel that I had been beguiled, in a sense, by Clive's easy, nonchalant, fluent conversation into thinking that he still had a great deal of general information at his disposal, despite the loss of memory for events. Given his intelligence, ingenuity, and humor, it was easy to think this on meeting him for the first time. But repeated conversations rapidly exposed the limits of his knowledge. It was indeed as Deborah wrote in her book, Clive "stuck to subjects he knew something about" and used these islands of knowledge as "stepping stones" in his conversation. Clearly, Clive's general knowledge, or semantic memory, was greatly affected, too—though not as catastrophically as his episodic memory.

Yet semantic memory of this sort, even if completely intact, is not of much use in the absence of explicit, episodic memory. Clive is safe enough in the confines of his residence, for instance, but he would be hopelessly lost if he were to go out alone. Lawrence Weiskrantz comments on the need for both sorts of memory in his 1997 book *Consciousness Lost and Found*:

> The amnesic patient can think about material in the immediate present. . . . He can also think about items in his semantic memory, his general knowledge. . . . But thinking for successful everyday adaptation requires not only factual knowledge, but the ability to recall it on the right occasion, to relate it to other occasions, indeed the ability to reminisce.

This uselessness of semantic memory unaccompanied by episodic memory is also brought out by Umberto Eco in his novel *The Mysterious Flame of Queen Loana*, in which the narrator, an antiquarian bookseller and polymath, is a man of Eco-like intelligence and erudition. Though amnesic from a stroke, he retains the poetry he has read, the many languages he knows, his encyclopedic memory of facts; but he is nonetheless helpless and disoriented (and recovers from this only because the effects of his stroke are transient).

It is similar, in a way, with Clive. His semantic memory, while of little help in organizing his life, does have a crucial social role: it allows him to engage in conversation (though it is occasionally more monologue than conversation). Thus, Deborah wrote, "he would string all his subjects together in a row, and the other person simply needed to nod or mumble." By moving rapidly from one thought to another, Clive managed to secure a sort of continuity, to hold the thread of consciousness and attention intact—albeit precariously, for the thoughts were held together, on the whole, by superficial associations. Clive's verbosity made him a little odd, a little too much at times, but it was highly adaptive—it enabled him to reenter the world of human discourse.

In the 1986 film, Deborah quoted Proust's description of Swann waking from a deep sleep, not knowing at first where he was, who he was, what he was. He had only "the most rudimentary sense of existence, such as may lurk and flicker in the depths of an animal's consciousness," until memory came back to him, "like a rope let down from heaven to draw me up out of the abyss of not-being, from which I could never have escaped by myself." This gave him back his personal consciousness and identity. No rope from Heaven, no autobiographical memory will ever come down in this way to Clive.

FROM THE START THERE HAVE BEEN, for Clive, two realities of immense importance. The first of these is Deborah, whose presence and love for him have made life tolerable, at least intermittently, in the twenty or more years since his illness. Clive's amnesia not only destroyed his ability to retain new memories; it deleted almost all of his earlier memories, including those of the years when he met and fell in love with Deborah. He told Deborah, when she questioned him, that he had never heard of John Lennon or John F. Kennedy. Though he always recognized his own children, Deborah told me, "he would be surprised at their height and amazed to hear he is a grandfather. He asked his younger son what O-level exams he was doing in 2005, more than twenty years after Edmund left school." Yet somehow he always recognized Deborah as his wife, when she visited, and felt moored by her presence, lost without her. He would rush to the door when he heard her voice, and embrace her with passionate, desperate fervor. Having no idea how long she had been away—since anything not in his immediate field of perception and attention would be lost, forgotten, within seconds—he seemed to feel that she, too, had been lost in the abyss of time, and so her "return" from the abyss seemed nothing short of miraculous. As Deborah put it:

> Clive was constantly surrounded by strangers in a strange place, with no knowledge of where he was or what had happened to him. To catch sight of me was always a massive relief—to

know that he was not alone, that I still cared, that I loved him, that I was there. Clive was terrified all the time. But I was his life, I was his lifeline. Every time he saw me, he would run to me, fall on me, sobbing, clinging.

How, why, when he recognized no one else with any consistency, did Clive recognize Deborah? There are clearly many sorts of memory, and emotional memory is one of the deepest and least understood.

The neuroscientist Neal J. Cohen recounts the famous story of Édouard Claparède, a Swiss physician who, upon shaking hands with a severely amnesic woman,

> pricked her finger with a pin hidden in his hand. Subsequently, whenever he again attempted to shake the patient's hand, she promptly withdrew it. When he questioned her about this behavior, she replied, "Isn't it allowed to withdraw one's hand?" and "Perhaps there is a pin hidden in your hand," and finally, "Sometimes pins are hidden in hands." Thus the patient learned the appropriate response based on previous experience, but she never seemed to attribute her behavior to the personal memory of some previously experienced event.

For Claparède's patient, some sort of memory of the pain, an implicit and emotional memory, persisted. It seems certain, likewise, that in the first two years of life, even though one retains no explicit memories (Freud called this infantile amnesia), deep emotional memories or associations are nevertheless being made in the limbic system and other regions of the brain where emotions are represented—and these emotional memories may determine one's behavior for a lifetime. A recent paper by Oliver Turnbull, Evangelos Zois, et al., in the journal *Neuro-Psychoanalysis*, has shown that patients with amnesia can form emotional transferences to an analyst, even though they retain no explicit memory of the analyst or their previous meetings. Nonetheless, a strong emotional bond begins to develop. Clive and Deborah were newly married at the time of his encephalitis, and deeply in love for a few years before that. His passionate relationship with her, a relationship that began before his encephalitis, and one that centers in part on their shared love for music, has engraved itself in him—in areas of his brain unaffected by the encephalitis—so deeply that his amnesia, the most severe amnesia ever recorded, cannot eradicate it.

Nonetheless, for many years he failed to recognize Deborah if she chanced to walk past, and even now he cannot say what she looks like unless he is actually looking at her. Her appearance, her voice, her scent, the way they behave with each other, and the intensity of their emotions and interactions—all this confirms her identity, and his own.

The other miracle was the discovery Deborah made early on, while Clive was still in the hospital, desperately confused and disoriented: that his musical powers were totally intact. "I picked up some music," Deborah wrote,

> and held it open for Clive to see. I started to sing one of the lines. He picked up the tenor lines and sang with me. A bar or so in, I suddenly realized what was happening. He could still read music. He was singing. His talk might be a jumble no one could understand but his brain was still capable of music. . . . When he got to the end of the line I hugged him and kissed him all over his face. . . . Clive could sit down at the organ and play with both hands on the keyboard, changing stops, and with his feet on the pedals, as if this were easier than riding a bicycle. Suddenly we had a place to be together, where we could create our own world away from the ward. Our friends came in to sing. I left a pile of music by the bed and visitors brought other pieces.

Miller's film showed dramatically the virtually perfect preservation of Clive's musical powers and memory. In these scenes from only a year or so after his illness, his face often appeared tight with torment and bewilderment. But when he was conducting his old choir, he performed with great sensitivity and grace, mouthing the melodies, turning to different singers and sections of the choir, cuing them, encouraging them, to bring out their special parts. It is obvious that Clive not only knew the piece intimately—how all the parts contributed to the unfolding of the musical thought—but also retained all the skills of conducting, his professional persona, and his own unique style.

Clive cannot retain any memory of passing events or experience and, in addition, has lost most of the memories of events and experiences *preceding* his encephalitis—how, then, does he retain his remarkable knowledge of music, his ability to sight-read, play the piano and organ, sing, and conduct a choir in the masterly way he did before he became ill?

H.M., a famous and unfortunate patient described by Scoville and Milner in 1957, was rendered amnesic by the surgical removal of both hippocampi, along with adjacent structures of the medial temporal lobes. (This was a desperate attempt at treating his intractable seizures; it was not yet realized that autobiographical memory and the ability to form new memories of events depended on these structures.) Yet H.M., though he lost many memories of his former life, did not lose any of the skills he had acquired, and indeed he could learn and perfect *new* skills with training and practice, even though he would retain no memory of the practice sessions.

Larry Squire, a neuroscientist who has spent a lifetime exploring mechanisms of memory and amnesia, emphasizes that no two cases of amnesia are the same. He wrote to me:

> If the damage is limited to the medial temporal lobe, then one expects an impairment such as H.M. had. With somewhat more extensive medial temporal lobe damage, one can expect something more severe, as in E.P. [a patient whom Squire and his colleagues have investigated intensively]. With the addition of frontal damage, perhaps one begins to understand Clive's impairment. Or perhaps one needs lateral temporal damage as

well, or basal forebrain damage. Clive's case is unique, because a particular pattern of anatomical damage occurred. His case is not like H.M. or like Claparède's patient. We cannot write about amnesia as if it were a single entity like mumps or measles.

Yet H.M.'s case and subsequent work made it clear that two very different sorts of memory could exist: a conscious memory of events (episodic memory) and an unconscious memory for procedures—and that such procedural memory is unimpaired in amnesia.

This is dramatically clear with Clive, too, for he can shave, shower, look after his grooming, and dress elegantly, with taste and style; he moves confidently and is fond of dancing. He talks abundantly, using a large vocabulary; he can read and write in several languages. He is good at calculation. He can make phone calls, and he can find the coffee things and find his way about the home. If he is asked how to do these things, he cannot say, but he does them. Whatever involves a sequence or pattern of action, he does fluently, unhesitatingly.

But can Clive's beautiful playing and singing, his masterly conducting, his powers of improvisation be adequately characterized as "skills" or "procedures"? For his playing is infused with intelligence and feeling, with a sensitive attunement to the musical structure, the composer's style and mind. Can any artistic or creative performance of this calibre be adequately explained by "procedural memory"? Episodic or explicit memory, we know, develops relatively late in childhood and is dependent on a complex brain system involving the hippocampi and medial temporal-lobe structures, the system that is compromised in severe amnesiacs and all but obliterated in Clive. The basis of procedural or implicit memory is less easy to define, but it certainly involves larger and more primitive parts of the brain—subcortical structures like the basal ganglia and cerebellum and their many connections to each other and to the cerebral cortex. The size and variety of these systems guarantee the robustness of procedural memory and the fact that, unlike episodic memory, procedural memory can remain largely intact even in the face of extensive damage to the hippocampi and medial temporal-lobe structures.

Episodic memory depends on the perception of particular and often unique events, and one's memories of such events, like one's original perception of them, are not only highly individual (colored by one's interests, concerns, and values) but prone to be revised or recategorized every time they are recalled. This is in fundamental contrast to procedural memory, where it is all-important that the remembering be literal, exact, and reproducible. Repetition and rehearsal, timing and sequence are of the essence here. Rodolfo Llinás, the neuroscientist, uses the term "fixed action pattern" (FAP) for such procedural memories. Some of these may be present even before birth (fetal horses, for example, may gallop in the womb). Much of the early motor development of the child depends on learning and refining such procedures, through play, imitation, trial and error, and incessant rehearsal. All of these start to develop long before the child can call on any explicit or episodic memories.

Is the concept of fixed action patterns any more illuminating than that of procedural memories in relation to the enormously complex, creative performances of a professional musician? In his book *I of the Vortex*, Llinás writes:

> When a soloist such as Heifetz plays with a symphony orchestra accompanying him, by convention the concerto is played purely from memory. Such playing implies that this highly specific motor pattern is stored somewhere and subsequently released at the time the curtain goes up.

But for a performer, Llinás writes, it is not sufficient to have implicit memory only; one must have explicit memory as well:

> Without intact explicit memory, Jascha Heifetz would not remember from day to day which piece he had chosen to work on

previously, or that he had ever worked on that piece before. Nor would he recall what he had accomplished the day before or by analysis of past experience what particular problems in execution should be a focus of today's practice session. In fact, it would not occur to him to have a practice session at all; without close direction from someone else he would be effectively incapable of undertaking the process of learning any new piece, irrespective of his considerable technical skills.

This, too, is very much the case with Clive, who, for all his musical powers, needs "close direction" from others. He needs someone to put the music before him, to get him into action, and to make sure that he learns and practices new pieces.

What is the relationship of action patterns and procedural memories, which are associated with relatively primitive portions of the nervous system, to consciousness and sensibility, which depend on the cerebral cortex? Practice involves conscious application, monitoring what one is doing, bringing all one's intelligence and sensibility and values to bear—even though what is so painfully and consciously acquired may then become automatic, coded in motor patterns at a subcortical level. Each time Clive sings or plays the piano or conducts a choir, automatism comes to his aid. But what happens in an artistic or creative performance, though it depends on automatisms, is anything but automatic. The actual performance reanimates him, engages him as a creative person; it becomes fresh and perhaps contains new improvisations or innovations. Once Clive starts playing, his "momentum," as Deborah writes, will keep him, and the piece, going. Deborah, herself a musician, expresses this very precisely:

The momentum of the music carried Clive from bar to bar. Within the structure of the piece, he was held, as if the staves were tramlines and there was only one way to go. He knew exactly where he was because in every phrase there is context implied, by rhythm, key, melody. It was marvellous to be free. When the music stopped Clive fell through to the lost place. But for those moments he was playing he seemed normal.

Clive's performance self seems, to those who know him, just as vivid and complete as it was before his illness. This mode of being, this self, is seemingly untouched by his amnesia, even though his autobiographical self, the self that depends on explicit, episodic memories, is virtually lost. The rope that is let down from Heaven for Clive comes not with recalling the past, as for Proust, but with performance—and it holds only as long as the performance lasts. Without performance, the thread is broken, and he is thrown back once again into the abyss.

Deborah speaks of the "momentum" of the music in its very structure. A piece of music is not a mere sequence of notes but a tightly organized organic whole. Every bar, every phrase arises organically from what preceded it and points to what will follow. Dynamism is built into the nature of melody. And over and above this there is the intentionality of the composer, the style, the order, and the logic that he has created to express his musical ideas and feelings. These, too, are present in every bar and phrase. Schopenhauer wrote of melody as having "significant intentional connection from beginning to end" and as "one thought from beginning to end." Marvin Minsky compares a sonata to a teacher or a lesson:

No one remembers, word for word, all that was said in any lecture, or played in any piece. But if you understood it once, you now own new networks of knowledge, about each theme and how it changes and relates to others. Thus, no one could remember Beethoven's Fifth Symphony entire, from a single hearing. But neither could one ever hear again those first four notes as just four notes! Once but a tiny scrap of sound; it is now a Known Thing—a locus in the web of all the other things we know, whose meanings and significances depend on one another.

A piece of music will draw one in, teach one about its structure and secrets, whether one is listening consciously or not. This is so even if one has never heard a piece of music before. Listening to music is not a passive process but intensely active, involving a stream of inferences, hypotheses, expectations, and anticipations. We can grasp a new piece—how it is constructed, where it is going, what will come next—with such accuracy that even after a few bars we may be able to hum or sing along with it. Such anticipation, such singing along, is possible because one has knowledge, largely implicit, of musical "rules" (how a cadence must resolve, for instance) and a familiarity with particular musical conventions (the form of a sonata, or the repetition of a theme). When we "remember" a melody, it plays in our mind; it becomes newly alive.

Thus we can listen again and again to a recording of a piece of music, a piece we know well, and yet it can seem as fresh, as new, as the first time we heard it. There is not a process of recalling, assembling, recategorizing, as when one attempts to reconstruct or remember an event or a scene from the past. We recall one tone at a time, and each tone entirely fills our consciousness yet simultaneously relates to the whole. It is similar when we walk or run or swim—we do so one step, one stroke at a time, yet each step or stroke is an integral part of the whole. Indeed, if we think of each note or step too consciously, we may lose the thread, the motor melody.

It may be that Clive, incapable of remembering or anticipating events because of his amnesia, is able to sing and play and conduct music because remembering music is not, in the usual sense, remembering at all. Remembering music, listening to it, or playing it, is wholly in the present. Victor Zuckerkandl, a philosopher of music, explored this paradox beautifully in 1956 in *Sound and Symbol*:

> The hearing of a melody is a hearing *with* the melody. . . . It is even a condition of hearing melody that the tone present at the moment should fill consciousness *entirely*, that *nothing* should be remembered, nothing except it or beside it be present in

consciousness. . . . Hearing a melody is hearing, having heard, and being about to hear, all at once. . . . Every melody declares to us that the past can be there without being remembered, the future without being foreknown.

It has been twenty years since Clive's illness, and, for him, nothing has moved on. One might say he is still in 1985 or, given his retrograde amnesia, in 1965. In some ways, he is not anywhere at all; he has dropped out of space and time altogether. He no longer has any inner narrative; he is not leading a life in the sense that the rest of us do. And yet one has only to see him at the keyboard or with Deborah to feel that, at such times, he is himself again and wholly alive. It is not the remembrance of things past, the "once" that Clive yearns for, or can ever achieve. It is the claiming, the filling, of the present, the now, and this is only possible when he is totally immersed in the successive moments of an act. It is the "now" that bridges the abyss.

As Deborah recently wrote to me, "Clive's at-homeness in music and in his love for me are where he transcends amnesia and finds continuum—not the linear fusion of moment after moment, nor based on any framework of autobiographical information, but where Clive, and any of us, *are* finally, where we are who we are."

BEN MCGRATH

Muscle Memory

FROM *THE NEW YORKER*

Artificial limbs have moved from primitive peg legs to sophisticated prostheses that mimic human movement, look increasingly lifelike, and are increasingly controlled by thought. Ben McGrath travels to the science-fiction-like frontier of the new prosthetics.

IN MAY OF 2004, a twenty-three-year-old former Marine named Claudia Mitchell went for a ride on the back of a friend's motorcycle along State Highway 71, in western Arkansas, near where she had grown up. Soon they were going faster than she was comfortable with, and she remembers feeling more alarmed than she had felt during the ferocious sandstorms she endured in Kuwait, when she couldn't see her fingers. As she and her friend approached

Mountainburg, she says, they took a sharp curve at high speed. The bike spun out of control. Mitchell was thrown, hit a guardrail, and came to rest, on her back, in a thicket. Her friend was unconscious. Mitchell felt extreme pain in her abdomen. (She later learned that her spleen had burst.) Volunteers from the local fire department arrived, and when an emergency medical technician found her she was desperately trying to extricate herself from the thicket.

"I kept saying, 'My arm isn't working! My arm isn't working!'" Mitchell recalled recently. "I was trying to push—it wasn't doing anything." Her arm—her left arm—wasn't working because it was over by the guardrail. It had been severed, just below the shoulder, on impact, although her brain continued to send signals to the primary motor nerves—the median, the ulnar, the radial, the musculocutaneous—that activate the muscles in the arm and hand. "I didn't understand why everybody kept saying there was something so seriously wrong," she went on. "I was in pain, but I didn't see any blood. The EMT said, 'Yeah, honey, it's gone, but we found it.'" He put the arm in an ice chest and brought it to the hospital.

Mitchell spent the next three days heavily sedated. (Her friend had sustained major injuries and was unconscious for weeks.) When she regained full consciousness, she learned that her doctors had decided against reattaching the arm, for fear of infection. As her recovery progressed, she resolved to learn to live with the loss of limb, but her involuntary nervous system resisted. A week after leaving the hospital, while crawling underneath a desk to get at some computer cables, Mitchell reached out to grab one with her right arm. But it was her only load-bearing arm, and she fell hard into the wall, stump first. The next three months were filled with similar mishaps. "I was constantly falling on my shoulder or reaching for things with my little shoulder, and it wasn't going anywhere," she said.

In *Moby-Dick*, Ahab complains of lingering pain in his missing leg, almost like a ghost, and the ship's carpenter remarks, "Yes, I have heard something curious on that score, sir; how that a dismasted man never entirely loses the feeling of his old spar, but it will be still pricking him at times." Mitchell, too, felt that somehow her arm was not so much gone as merely invisible—a phenomenon known as phantom limb. (Lord Nelson, having lost his right arm to cannon fire in 1797, considered his phantom fingers to be proof of the existence of a soul.) When Mitchell was fitted with a mechanical prosthesis, she tended to keep it bent at a ninety-degree angle, as though in a sling, giving visual representation to the posture of her throbbing phantom, which she was powerless to move.

Mitchell's prosthesis was a state-of-the-art battery-powered robotic arm that operated myoelectrically; that is, by using electrodes to amplify the electrical charges from muscle contractions and drive a motorized elbow or hand. It was heavy, however, and she found it slow, and cumbersome—nearly useless. Before long, she stopped wearing it, and learned to tie her shoes and to type using one hand. She got by. Her friends called her the Queen of Backspace.

THE ROBOTIZATION OF HUMANS for medical purposes is in some respects already highly advanced. Cochlear implants replicate hearing through the electrical stimulation of auditory nerves, artificial retinas promise to undo the effects of blindness, and even automated bladder control for the incontinent is now available, at least in laboratory prototype. Medical researchers have begun to explore the possibility that people can regenerate lost appendages, in the manner of salamanders and starfish, and are harvesting extracellular powder from pigs' bladders, which may prove useful in growing new human-finger tissue. But when it comes to real locomotive hardware—the stuff of Darth Vader, functionally speaking—we're still closer to Captain Hook than to RoboCop.

References to an artificial limb can be found in the Rigveda, an ancient Sanskrit text dating to around 1400 B.C., and later in Herodotus, but the earliest prosthesis yet recovered belonged to the city of Capua: a leg from about 300 B.C., made of bronze, iron, and wood. Metal reinforcements gave way, in time, to other materials, including whalebone (Ahab's replacement spar), but it wasn't until after the Civil War, when amputees on the Union side alone numbered thirty thousand, and medical advances were at last sufficient to keep nearly three-quarters of them alive, that an organized prosthetics industry arose, with a more or less standardized means of, for instance, mimicking arm function. Wooden hands, harnessed to leather sockets, were controlled by straps running through a system of eyelets and pulleys that fit around the shoulder and back. The engineering was far from efficient, and these systems suffered from high frictional losses.

The inventor Frank Bowden, who founded the Raleigh Bicycle Company in 1888, took the arrangement to the next step. While playing around with alternatives to backpedalling as a means of slowing down, he devised hand brakes that transmitted force to the wheel rims through a set of stainless-steel wires sheathed in hollow tubes. These Bowden cables, as they are called today, offered the advantage of graduated feedback, affording riders greater control over their speed. Bowden cables were subsequently used in early airplanes, for controlling wing flaps, where such feedback is crucial. After the Second World War, Northrop Corporation, the aviation pioneer, put its aeronautical engineers to work on prosthetics and became an early builder of cable-operated arm systems. Cables replaced leather straps, and could be used reasonably well to manipulate split hooks, which provide a crude sort of opposable thumb.

Around the same time, in Munich, a physicist named Reinhold Reiter discovered myoelectric control, and, using vacuum tubes for amplification, created a stationary prototype electric hand that could be opened and closed through contractions of the residual muscles. In the late 1950s, shortly after the launching of Sputnik, Soviet scientists successfully applied myoelectricity to the operation of a portable hand, which they introduced at the 1958 World Expo, in Brussels. But progress on the merger of man with machine has been slow, and modern myoelectric devices aren't really any better than the cable-and-pulley system. At the Walter Reed Army Medical Center, many amputees still opt for a cosmetic device—a nice, silicone hand glove, for example, attached to an aluminum shell of an arm—that has no functional capability.

Since the campaigns in Afghanistan and Iraq began, more than six hundred soldiers have returned home without arms or legs, thanks, in part, to modern body armor, which saves lives that would in earlier wars have been lost. Through the Defense Advanced Research Projects Agency, or DARPA, the Pentagon is funding an initiative called Revolutionizing Prosthetics, whose goal is to produce fully humanlike replacement arms. Fifty million dollars is to be spent on complementary research teams from the academy, led by Johns Hopkins, and the private sector—DEKA Research, founded by Dean Kamen, the inventor of the Segway.

"Right now, we've got a hook that opens and closes," Colonel Geoffrey Ling, a military neurologist who is overseeing the project, said recently. "That allows you to pick up a pencil, but it doesn't allow you to write. You can pick up a fork, but it doesn't allow you to feed yourself." DEKA hopes to unveil an eight-pound mechanical arm—which is lighter than a human arm—before the end of the year, and in April the Johns Hopkins researchers produced Proto 1, which offers the advantage, among other things, of swinging freely at the shoulder and the elbow, thereby allowing for more natural walking. DARPA's plan is to have amputee soldiers not only eating, writing, and walking comfortably but cleaning rifles and throwing grenades within two years.

WHEN CLAUDIA MITCHELL TRIED to push her way out of the thicket beside Highway 71, her brain repeatedly sent commands along nerves that now went nowhere. Dr. Todd Kuiken, a professor of biomedical engineering at Northwestern University and the director of the Neural Engineering Center for Bionic Medicine at the Rehabilitation Institute of Chicago, likens her situation to that of an unplugged telephone. "So you and I are talking on the phone," he says. "And that phone cable is like a nerve. I talk—that's my motor command. And I listen to you—that's my sensory coming back through the phone cable. Let's say your phone breaks, or you have an amputation. My voice is still going to that cable, and if it was hooked up to something you'd be able to hear it."

Conventional myoelectric technology ignores the disconnected nerves and attempts to harness healthy but remote muscle signals (in the shoulder stump, say, or in the back) to power computerized stand-ins for a missing elbow or hand. For the amputee, this means learning a new, much clumsier language of movement, like Morse code as a substitute for direct speech: making a fist may now require the deliberate, if counterintuitive, contraction of one's upper deltoid, and bending an elbow may involve a sequence of shrugging motions in the shoulder and neck. The burden is cognitive rather than physical, but the system is hardly more efficient than leather straps and pulleys.

If the brain can persist in thinking, on some level, that it still has a left arm to push off with, Kuiken reasons, why not give those thoughts a muscle to flex? Any muscle ought to do, as long as you can program a computer to articulate its contractions. "It's like if you go and buy a new phone and plug it into the old cord, and call me up," he says. "We can talk again, and I don't know you have a new phone."

As a teenager in the 1970s, Kuiken was a self-described gearhead and a science-fiction fan who watched *The Six Million Dollar Man* on TV. For years, Kuiken experimented with nerve transfers on rats, but it wasn't until 2002 that he got his first chance to put his idea into practice with a human. That winter, he operated on Jesse Sullivan, a fifty-five-year-old power-company technician who had lost both arms after touching a live wire. Gregory Dumanian, a surgeon on Kuiken's team, rerouted the ulnar, radial, median, and musculocutaneous nerves from Sullivan's left-shoulder stump to his pectoral muscle, which he then carved into four smaller pieces, one for each nerve. (The pectoral muscle was of no use to Sullivan in its original form, because he lacked arms.) With the brain-muscle connection reestablished, Sullivan was eventually able to operate a myoelectric prosthesis using nothing other than intuition. Just thinking about bending his left elbow activated the mechanical elbow socket. In a video, shot six months after the procedure, Sullivan can be seen moving his extended left arm from side to side and saying, "Wax on, wax off," in imitation of the Karate Kid. (For the right arm, he continues to use a cable and a hook.)

In the fall of 2004, a friend of Claudia Mitchell's saw a brief mention of Sullivan's success story in *Popular Science*. The following spring, Mitchell got in touch with Kuiken, and that August he and his team at the Rehabilitation Institute performed a similar procedure on her. In addition, they severed the existing sensory nerves in a small patch of her chest, creating a numb spot roughly four inches wide above the left breast. The relocated arm nerves then grew into this numbed patch of surface skin, reinnervating it. The results were revealed last September at a press conference in Washington, D.C., where the assembled science reporters proclaimed Mitchell the world's first "bionic woman." On a makeshift stage, she and Jesse Sullivan exchanged intuitive robotic high fives. Not every stunt went according to plan: Mitchell dropped a bottle of water as she tried to raise it to her lips.

I met Mitchell and Kuiken in New York five months later, and she said that it took her a half hour just to figure out how to tie her shoes with two hands again, but that, once she recalled the proper order of operations, she was able to do it without great difficulty. "We've rewired Claudia," Kuiken explained. "We're rewiring a human to work with a machine."

Mitchell, who is small and unassuming and has long dark hair, was wearing a loose-fitting pink sweater and kept her arms folded across her lap, her right hand gently squeezing her left. The nails had been painted identically on both hands, and she wore a bracelet and a pinkie ring on her left—which, on closer inspection, turned out to be rubber (concealing metal). Her posture was awkwardly stiff, and her left shoulder appeared broader than her right, because of the harness required to hold the prosthesis in place. At Kuiken's suggestion, she opened and closed both hands, simultaneously, and then began recoiling her arms at the elbows, as if she were lifting barbells.

Kuiken is one of the scientists involved with the Revolutionizing Prosthetics project, but Mitchell's prosthesis, he explained, was not a futuristic DARPA special; it was a commercially available model that cost sixty thousand dollars, and its limitations soon became evident. It can lift only ten pounds. Two of its motors (the hand and the elbow) operated via "thought control," as Kuiken calls it, while a third, a wrist rotator, required Mitchell to manipulate a small switch near her shoulder. The motors, moreover, were noisy, and the degree of freedom in her arm was essentially confined to three types of movement. She could open and close her hand, she could flex and extend her elbow—and she could do these simultaneously, rather than sequentially, as most amputees are forced to do—but little else was possible. Kuiken named a series of more specialized hand movements—fine pinch, trigger, key grip, three-jaw chuck, power grasp—which he hoped to teach Mitchell in the coming months. She demonstrated each using her human hand. For the three-jaw chuck, this meant employing the thumb, forefinger, and middle finger in the manner of a vending-machine claw used to retrieve stuffed animals.

In the interest of increasing Mitchell's freedom of movement, Kuiken has developed a prototype of what he calls a "six-motor arm," which he keeps at his lab in Chicago. He assembled it in patchwork fashion, like a Sunday garage mechanic, using existing components gathered from around the world. It is composed of a "Boston digital elbow" (long the industry standard), a Scottish-made shoulder, a German wrist rotator, a Chinese hand-and-wrist flexor, and a humeral rotator that was made in-house, by Richard Weir, a colleague at Northwestern. "We're like the Wright brothers here," Kuiken said. "This is the first time it's flown, but we're already getting to that biplane."

WHEN MITCHELL SHOWERS, THE water hitting the patch of reinnervated skin on her chest makes her feel as though her nonexistent left arm were getting wet. "When we wake the sensation in the arm back up, it's not a question of 'what I used to call my arm,'" Paul Marasco, a postdoctoral fellow at the Rehabilitation Institute, said in April. "It's, like, 'My arm's *back*.'" It simply lives in her chest now.

Marasco was looking at a peculiar sort of anatomical map, which showed Mitchell's chest. The map had been made, in the manner of all the early explorers' maps, by trial and error—pushing on Mitchell's chest with Q-tips, asking what she felt, and recording her description. It showed, within the chest, an array of hands, each shaded to indicate a more localized area of the palm or of specific fingers. A simplified diagram showed her overall hand region—the original numb spot—laid out in a seventy-two-point grid, with her pinkie, for example, primarily occupying a portion of the lower right-hand quadrant, and her thumb just left of center.

Mitchell had deposited her prosthesis on a shelf that served as a kind of parking garage for artificial limbs (hers was identifiable by the jewelry), and removed her shirt, a light-green button-down tied off at the sleeve, in favor of what she called a toga. Marasco presented her with a perforated thermoplastic casing that fit snugly over her shoulder and chest. In addition to the perforations, which represented the seventy-two grid points, the exterior of the casing had been marked in a couple of places to indicate a scar and a mole on Mitchell's skin. Marasco aligned the markings with the actual blemishes underneath, and then, using a Sharpie, made a dot on her chest through hole No. 21, which, according to the map, represented a sensation of compression on the inside edge of her thumb and the outside edge of her index finger.

Mitchell settled into a reclining chair, similar to one in a dentist's office, and put on a pair of headphones, which she had plugged into an iPod. Aimee Schultz, a lab engineer, positioned a different kind of mechanical arm—a metal bar, with knobs and a metal probe for a hand—above Mitchell's chest, in preparation for a series of vibration tests. These were intended to compare the sensitivity of her reinnervated skin with her normal skin—her virtual hand with her real one. Currently, Mitchell receives no sensory feedback from her prosthesis. (In New York, when I asked her to squeeze a can of Coke using her artificial hand, with her eyes closed, she was unable to tell me whether, or at what point, she had succeeded.) But Kuiken's team is working to develop small touch pads that will be added to her fingers and palm for the purpose of sending signals pertaining to temperature, force, and texture back to the corresponding locations on her chest map. The pads must be carefully calibrated. Early testing has suggested, for instance, that Mitchell's bionic side is slightly more tolerant of extremely cold temperatures; the pads could be adjusted to take that into account.

The vibration tests were set up like an educational video game. Every few seconds, the words "Applying Stimulus" would appear on a screen in front of Mitchell, and the probe, fixed on the Sharpie dot, either would or would not vibrate, at varying amplitudes and frequencies. Her assignment was to indicate whether or not she felt any vibration by using a mouse. It was a painstaking and tedious assignment, and to keep her motivated a counter on the bottom tallied her "score."

AFTER HIGH SCHOOL, MITCHELL worked in a meat-processing plant for a few months, cutting giant slabs of beef that were then shipped to various restaurant franchises: Outback, Steak and Ale. "I said to myself, 'There has got to be something better than this,'" she recalled. She decided to join the Marines. Now twenty-six, she is still active in Marine causes—she leads a Young Marines unit for preteens near her home, in Maryland, and volunteers with the Marines Helping Marines program—but she is also a full-time student, having just finished her first year at Howard Community College, where she's majoring in communications. She visits Kuiken's lab in Chicago during school vacations, as she did in April; there, her "occupational therapy" training sessions seem to have been culled from a home-economics textbook written in 1953. She cooks, irons shirts, and tosses salads—with and without the prosthesis—for practice. (Lab staffers, who get to join in the tasting component of her evaluation, remember her fried okra and spice cake with special fondness.) During one baking session last year, Mitchell was using her prosthesis to steady a bowl while holding an electric mixer in her right hand. When she turned on the mixer, the current somehow caused her prosthesis to extend without warning, and it shook the bowl, spilling the contents. "I started making one of the doctors nervous," she said recently. "I was, like, 'You want to see my new trick?'"

The potential for more useful robo-tricks long ago occurred to Mitchell, and when she returned from lunch on the day of the vibration tests she reminded Blair Lock, another lab engineer, that she wanted her next arm to be equipped with an embedded iPod, as well as a Palm Pilot. "Be careful, we might give you a GPS tracker, too," Lock said. "We'll know your every move." Lock was more concerned, in the interim, with teaching the arm's computer processor to recognize more subtle motor commands than the three types of locomotion Mitchell currently enjoyed, and he had plugged a laptop into a giant flat-screen monitor with the afternoon's agenda—advanced signal processing—in mind. "Is that new?" Mitchell asked, eying the monitor. "I love DARPA!"

Lock affixed another plastic casing to Mitchell's shoulder and chest, this time with thirteen electrodes and a bundle of wires on the underside. The wires were connected to the laptop, and, once the electrodes had been properly aligned with Mitchell's chest muscles, a rainbow-colored electromyogram, or EMG, depicting the electrical activity in the muscles, appeared on the monitor. Lock then played the part of physical therapist or personal trainer. "Now elbow up all the way," he said. "Down all the way. Relax." Mitchell sat motionless, on the edge of a table, but on the monitor a video image of a man performed the respective actions almost immediately, while the laptop recorded the digital pattern on the EMG which was associated with each intended movement. "Wrist flexion in all the way," Lock continued. "Good. Back to neutral. Now, hand closed. Leave the hand closed and extend the wrist. . . . Now, how about a supination. Pronate?" The virtual man's forearm turned over. "Back and forth with those. Hand closed again. OK, relax. Am I tiring you out?"

Mitchell nodded. She'd remained perfectly still, but the figure on the screen had been busily engaged in what appeared to be a nonstop slow-motion dance workout. Mitchell's chest muscles were becoming sore from overwork.

The exercises continued—there were about a dozen discrete arm motions in this new repertoire—with only one minor hitch, in the case of humeral rotation, when Mitchell's phantom-limb sensation, which has persisted in spite of the reinnervation surgery, seemed to interfere. "You know how my favorite place in the world for my arm is like this, because that's pretty much how it feels?" she said, making an L shape with her right arm against her stomach to illustrate. "So rotation out is no big deal"—she swung her hand away from her body—"but rotation in is weird, because it's not that big of a movement." For the virtual man on the screen, who lacked legs or a waist, this was not a problem.

"Blair, am I really going to have an arm that'll do all this?" Mitchell asked, sounding excited, after they had finished. "I want my arm to have muscles. Make it a little bit poofy here." She made a flexed-biceps motion with her right arm. "That'll give you a little more room to put wires in, anyway." She said she hoped that they'd make the arm capable of lifting a hundred and fifty pounds, so that she could impress her guy friends at the gym. She was sweating—the casing is heavy, and takes a toll on one's posture. A staffer at the institute who ordinarily deals with wheelchair upholstery had been working on sewing the electrodes into a customfitted Spandex shirt for future use.

COLONEL LING, THE SUPERVISOR of the DARPA project, is interested in surgical techniques that involve implanting electrodes or computer chips underneath the skin, whether at the periphery (the nerve endings in the stump) or, better yet, in the motor cortex itself. "We need a closed loop that really involves the brain," he said, explaining that proprioception, or awareness of the body's movement through space, is a crucial component of fully human arm function, and that only direct sensory feedback to the brain is capable of restoring it. Typing and playing the piano without looking at the keys, for instance, require proprioception.

Scientists funded by DARPA have been experimenting on monkeys for years, using chips to power robotic arms. "We've got signals from monkeys right now that can drive a keyboard," Ling told me recently. "We have video of monkeys, actually controlling arms, working in 3-D space." He added, "I know we can do it. At what point do we put them into the human, you might ask? There are always human-use issues, and so on and so forth. If there were no such thing as human-use issues, we would be into humans dramatically sooner. But we have to follow ethics. We have to do the right thing." He was confident the right thing could be done quickly. "My hope is that we're going to be putting it into humans within the next eighteen to twenty months. Maybe I'm being bold, but, hey, opportunity favors the bold."

There are, in fact, humans suffering from severe paralysis who have already benefited from motor-cortex surgery; one research group working toward this goal refers to its implantable sensor as a "neuromotor prosthesis." But where otherwise healthy amputees are concerned, Kuiken, for his part, favors the clinically practical—or what he calls the "flying-kid test," which he thought of one day while playing with his two-year-old son. "He comes in and stomps on me," he explained. "I get bruises in places I never imagined. But at least they heal." He is reluctant to recommend for a young and active patient any procedure that could be complicated or undone—chips dislodged, say—in the course of ordinary roughhousing.

"I told Doc he can do whatever he wants to my arm, but he ain't messing with my head," Mitchell said recently, reinforcing Kuiken's instinct. Since last fall, she has run in two marathons, and in April she mentioned that most of her time outside school was spent attending to the needs of her boyfriend, who was shot in Iraq last November, in an unprotected area between the bottom of his flak jacket and his pelvic brace. Historically, the demands and rigors of war may have provided the impetus for scientific and medical progress, but domestic life is often where the significant advancements are felt first. In Mitchell's case, she said that occasionally she now finds herself reclining on the sofa at home, watching television, and will suddenly realize that the arm propping her head up is her left arm.

MARGARET TALBOT

Duped

FROM *THE NEW YORKER*

With new medical brain-scanning technologies becoming more pow-erful, the race is on for a new kind of lie detector. Margaret Talbot investigates whether these new devices work.

THE MOST EGREGIOUS LIAR I ever knew was someone I never suspected until the day that, suddenly and irrevocably, I did. Twelve years ago, a young man named Stephen Glass began writing for *The New Republic*, where I was an editor. He quickly established himself as someone who was always onto an amusingly outlandish story—like the time he met some Young Republican types at a convention, gathered them around a hotel-room minibar, then, with guileless ferocity, captured their boorishness in print. I liked Steve; most of us who worked with him did. A baby-faced guy from suburban Chicago, he padded around the office in his socks. Before going on an errand, Steve would ask if I wanted a muffin or a sandwich; he always noticed a new scarf or a clever turn of phrase, and asked after a colleague's baby or spouse. When he met with editors to talk about his latest reporting triumph, he was self-effacing and sincere. He'd look us in the eye, wait for us to press him for details, and then, without fidgeting or mumbling, supply them.

One day, the magazine published an article by Steve about a teenager so diabolically gifted at hacking into corporate computer networks that CEOs paid him huge sums just to stop messing with them. A reporter for the online edition of *Forbes* was assigned to chase down the story. You can see how Steve's journalism career unravelled if you watch the movie *Shattered Glass*: *Forbes* challenged the story's veracity, and Steve—after denying the charges, concocting a fake Web site, and enlisting his brother to pose as a victimized CEO—finally confessed that he'd made up the whole thing. Editors and reporters at the magazine investigated, and found that Steve had been inventing stories for at least a year. The magazine disavowed twenty-seven articles.

After Steve's unmasking, my colleagues and I felt ashamed of our gullibility. But maybe we shouldn't have. Human beings are terrible lie detectors. In academic studies, subjects asked to distinguish truth from lies answer correctly, on average, 54 percent of the time. They are better at guessing when they are being told the truth than when they are being lied to, accurately classifying only 47 percent of lies, according to a recent meta-analysis of some two hundred deception studies, published by Bella DePaulo, of the University of California, Santa Barbara, and Charles Bond, Jr., of Texas Christian University. Subjects are often led astray by an erroneous sense of how a liar behaves. "People hold a stereotype of the liar—as tormented, anxious, and conscience-stricken," DePaulo and Bond write. (The idea that a liar's anxiety will inevitably become manifest can be found as far back as the ancient Greeks, Demosthenes in particular.) In fact, many liars experience what deception researchers call "duping delight."

Aldert Vrij, a psychologist at the University of Portsmouth, in England, argues that there is no such thing as "typical" deceptive behavior—"nothing as obvious as Pinocchio's growing nose." When people tell complicated lies, they frequently pause longer and more often, and speak more slowly; but if the lie is simple, or highly polished, they tend to do the opposite. Clumsy deceivers are sometimes visibly agitated, but, over all, liars are less likely to blink, to move their hands and feet, or to make elaborate gestures—perhaps they deliberately inhibit their movements. As DePaulo says, "To be a good liar, you don't need to know what behaviors really separate liars from truthtellers, but what behaviors people *think* separate them."

A liar's testimony is often more persuasive than a truthteller's. Liars are more likely to tell a story in chronological order, whereas honest people often present accounts in an improvised jumble. Similarly, according to DePaulo and Bond, subjects who spontaneously corrected themselves, or said that there were details that they couldn't recall, were more likely to be truthful than those who did not—though, in the real world, memory lapses arouse suspicion.

People who are afraid of being disbelieved, even when they are telling the truth, may well look more nervous than people who are lying. This is bad news for the falsely accused, especially given that influential manuals of interrogation reinforce the myth of the twitchy liar. *Criminal Interrogation and Confessions* (1986), by Fred Inbau, John Reid, and Joseph Buckley, claims that shifts in posture and nervous "grooming gestures," such as "straightening hair" and "picking lint from clothing," often signal lying. David Zulawski and Douglas Wicklander's *Practical Aspects of Interview and Interrogation* (1992) asserts that a liar's movements tend to be "jerky and abrupt" and his hands "cold and clammy." Bunching Kleenex in a sweaty hand is another damning sign—one more reason for a sweaty-palmed, Kleenex-bunching person like me to hope that she's never interrogated.

Maureen O'Sullivan, a deception researcher at the University of San Francisco, studies why humans are so bad at recognizing lies. Many people, she says, base assessments of truthfulness on irrelevant factors, such as personality or appearance. "Baby-faced, non-weird, and extroverted people are more likely to be judged truthful," she says. (Maybe this explains my trust in Steve Glass.) People are also blinkered by the "truthfulness bias": the vast majority of questions we ask of other people—the time, the price of the breakfast special— are answered honestly, and truth is therefore our default expectation. Then, there's the "learning-curve problem." We don't have a refined idea of what a successful lie looks and sounds like, since we almost never receive feedback on the fibs that we've been told; the coworker who, at the corporate retreat, assured you that she loved your presentation doesn't usually reveal later that she hated it. As O'Sullivan puts it, "By definition, the most convincing lies go unde-tected."

MAYBE IT'S BECAUSE WE'RE such poor lie detectors that we have kept alive the dream of a foolproof lie-detecting machine. This February, at a conference on deception research, in Cambridge, Massachusetts, Steven Hyman, a psychiatrist and the provost of Harvard, spoke of "the incredible hunger to have some test that separates truth from deception—in some sense, the science be damned."

This hunger has kept the polygraph, for example, in widespread use. The federal government still performs tens of thousands of polygraph tests a year—even though an exhaustive 2003 National Academy of Sciences report concluded that research on the polygraph's efficacy was inadequate, and that when it was used to investigate a specific incident after the fact it performed "well above chance, though well below perfection." Polygraph advocates cite accuracy estimates of 90 percent—which sounds impressive until you think of the people whose lives might be ruined by a machine that fails one out of ten times. The polygraph was judged thoroughly unreliable as a screening tool; its accuracy in "distinguishing actual or potential security violators from innocent test takers" was deemed "insufficient to justify reliance on its use." And its success in criminal investigations can be credited, in no small part, to the intimidation factor. People who believe that they are in the presence of an infallible machine sometimes confess, and this is counted as an achievement of the polygraph. (According to law-enforcement lore, the police have used copy machines in much the same way: They tell a suspect to place his hand on a "truth machine"—a copier in which the paper has "lie" printed on it. When the photocopy emerges, it shows the suspect's hand with "lie" stamped on it.)

Over the past two decades, inventors have attempted to supplant the polygraph with new technologies: voice-stress analysis; thermal imaging of the face; and, most recently and spectacularly, brain imaging. Though these methods remain in an embryonic stage of development, they have already been greeted with considerable enthusiasm, especially in America. Private companies are eager to replace traditional modes of ascertaining the truth—such as the jury system—with a machine that can be patented and sold. And law-enforcement agencies yearn to overcome the problem of suspects who often remain maddeningly opaque, even in the face of sustained interrogation. Although one immediate result of the September 11th attacks was the revival of an older, and even more controversial, form of interrogation—torture—the war on terror has also inflamed the desire for a mind-reading machine.

Not long ago, I met with an entrepreneur named Joel Huizenga, who has started a company, based in San Diego, called No Lie MRI. Most methods of lie detection look at the activity of the sympathetic nervous system. The polygraph, for instance, is essentially an instrument for measuring stress. Heart and respiration rates, blood volume, and galvanic skin response—a proxy for palm sweat—are represented as tracings on graph paper or on a screen, which fluctuate with every heartbeat or breath. The method that Huizenga is marketing, which employs a form of body scanning known as functional magnetic resonance imaging, or fMRI, promises to look inside the brain. "Once you jump behind the skull, there's no hiding," Huizenga told me.

Functional MRI technology, invented in the early '90s, has been used primarily as a diagnostic tool for identifying neurological disorders and for mapping the brain. Unlike MRIs, which capture a static image, an fMRI makes a series of scans that show changes in the flow of oxygenated blood preceding neural events. The brain needs oxygen to perform mental tasks, so a rise in the level of oxygenated blood in one part of the brain can indicate cognitive activity there. (Blood has different magnetic properties when it is oxygenated, which is why it is helpful to have a machine that is essentially a big magnet.) Brain-scan lie detection is predicated on the idea that lying requires more cognitive effort, and therefore more oxygenated blood, than truthtelling.

Brain scanning promises to show us directly what the polygraph showed us obliquely. Huizenga expects his company to be a force for justice, exonerating customers who are, as he put it, "good people trying to push back the cruel world that is indicting them unfairly." Brain scans already have clout in the courtroom; during death-penalty hearings, judges often allow images suggesting neurological impairment to be introduced as mitigating evidence. In theory, an improved method of lie detection could have as profound an impact as DNA evidence, which has freed more than a hundred wrongly accused people since its introduction, in the late '80s. If Huizenga has perfected such a technology, he's onto something big.

At Huizenga's suggestion, we met at a restaurant called the Rusty Pelican, on the Pacific Coast Highway, in Newport Beach. A television screen on one wall showed a surfing contest; Huizenga, who is fifty-three, with dirty-blond hair in a boyish cut, is a surfer himself. He has a bachelor's degree from the University of Colorado, a master's degree in biology from Stony Brook, and an MBA from the University of Rochester. No Lie is Huizenga's second start-up. The first, ISCHEM Corporation, uses body scanning to look for plaque in people's arteries. Before that, he worked for Pantox, a company that offers blood tests to gauge a person's antioxidant levels.

After we sat down, Huizenga recounted the origins of No Lie. A few years ago, he came across an item in the *Times* about some tantalizing research conducted by Daniel Langleben, a psychiatrist and neuroscientist at the University of Pennsylvania. Subjects were placed inside an fMRI machine and told to make some true statements and some false ones. Brain scans taken while the subjects were lying frequently showed a significantly increased level of activity in three discrete areas of the cerebral cortex. Langleben suggested that "intentional deception" could be "anatomically localized" by fMRI scanning. Huizenga immediately saw a business opportunity. "I jumped on it," he told me. "If I wasn't here sitting in front of you, somebody else would be."

The Web site for No Lie claims that its technology, which is based on the Penn protocol, "represents the first and only direct measure of truth verification and lie detection in human history!" No Lie just started offering tests commercially, and has charged about a dozen clients approximately ten thousand dollars apiece for an examination. (No Lie sends customers to an independent imaging center in Tarzana, a suburb of Los Angeles, to insure that "quality testing occurs according to standardized test protocols.") Some of these initial clients are involved in civil and criminal cases; the first person to use the service, Harvey Nathan, was accused in 2003 of deliberately setting fire to a deli that he owns in South Carolina. A judge dismissed the charges, but Nathan wanted to bring suit against his insurance company, and he thought that documented evidence of his innocence would further his cause. So in December he flew to California and took No Lie's test. He passed. Nathan said, "If I hadn't, I would have jumped from the seventeenth floor of the hotel where I was staying. How could I have gone back to South Carolina and said, 'Oh that machine must not have worked right'? I believed in it then and I believe in it now." Nathan's exam was filmed for the

Discovery Channel, which may soon launch a reality series center-ing on brain-scanning lie detection.

Several companies have expressed interest in No Lie's services, Huizenga told me. (He would not name them.) He said that he will be able to accommodate corporate clients once he has signed deals with other scanning facilities; he is in talks with imaging centers in a dozen cities, including New York and Chicago. No Lie also plans to open a branch in Switzerland later this year.

Huizenga has been criticized for his company's name, but he said, "It's not about being dignified—it's about being remembered." He believes that the market for fMRI-based lie detection will one day exceed that of the polygraph industry, which brings in hundreds of millions of dollars annually. Investment analysts say that it is too soon to judge if Huizenga's optimism is warranted, but No Lie has attracted some prominent backing. One of its prime investors is Alex Hart, the former CEO of MasterCard International, who is also serving as a management consultant. And it has a "scientific board" consisting of four paid advisers, among them Terrence Sejnowski, the director of the Crick-Jacobs Center for Theoretical and Compu-tational Biology at the Salk Institute. In an e-mail, Sejnowski ex-plained that he offers counsel on "advanced signal processing and machine-learning techniques that can help improve the analysis of the data and the accuracy of the performance." He said of No Lie, "The demand is there, and to succeed as a company the new technol-ogy only needs to be better than existing approaches."

Huizenga speaks of his company's goals in blunt terms. "What do people lie about?" he asked me. "Sex, power, and money—probably in that order." (The company's Web site recommends No Lie's ser-vices for "risk reduction in dating," "trust issues in interpersonal relationships," and "issues concerning the underlying topics of sex, power, and money.") "Parents say, 'Yes, this is *perfect* for adoles-cents,'" he went on. "People who are dating say, 'Yes, this is great for dating, because people never tell you the truth.'"

He said that his company receives dozens of inquiries a week: from divorcing men accused of child abuse; from women wanting to prove their fidelity to jealous spouses or boyfriends; from people representing governments in Africa and the former Soviet republics; from "the Chinese police department." He said that he understood why governments were interested in lie-detection technology. "Look at Joe Stalin," he said. "Joe wanted power, he wanted to be on top. Well, it's hard to murder massive numbers of opponents. People in our government, and in others', need more effective ways of weeding out those who aren't their puppets." Some potential foreign clients had explained to him, he said, that in societies that lacked "civilization, there is no trust, and lie detection could help build that trust." (He wasn't sure about that—he was "mulling it over.") Huizenga said that the United States government was "interested" in the kind of technology offered by No Lie; the company has hired Joel S. Lisker, a former FBI agent, to be its "sales liaison for the federal government." (Lisker declined to be interviewed, saying that his government contacts were "confidential.")

The Pentagon has supported research into high-tech lie detection, including the use of fMRI. The major scientific papers in the field were funded, in part, by the Defense Advanced Research Projects Agency, which develops new technologies for military use, and by the Department of Defense Polygraph Institute, which trains lie-detection experts at Fort Jackson, South Carolina. (The Polygraph Institute underwent a name change in January—it's now the Defense Academy for Credibility Assessment—apparently in deference to new technologies such as fMRI.) Last June, the ACLU filed several Freedom of Information Act requests in an attempt to learn more about the government's involvement with the technology. Chris Calabrese, an ACLU lawyer, said that the C.I.A. would neither "confirm nor deny" that it is investigating fMRI applications; the Pentagon produced PowerPoint presentations identifying brain scans as a promising new technology for lie detection. Calabrese went on, "We were motivated by the fact that there are companies trying to sell this technology to the government. This administration has a history of using question-able techniques of truth verification."

Many scholars also think that Huizenga's effort is premature. Steven Hyman, the Harvard professor, told me that No Lie was "fool-ish." But the history of lie-detection machines suggests that it would be equally foolish to assume that a few scholarly critics can forestall the adoption of such a seductive new technology. "People are drawn to it," Huizenga said, smiling. "It's a magnetic concept."

IN COMIC BOOKS OF the 1940s, Wonder Woman, the sexy Ama-zon superhero, wields a golden "lasso of truth." Anybody she cap-tures is rendered temporarily incapable of lying. Like the golden lasso, the polygraph, its inventors believed, compelled the body to reveal the mind's secrets. But the connection between the lasso and the lie detector is even more direct than that: Wonder Woman's cre-ator, William Moulton Marston, was also a key figure in the develop-ment of the polygraph. Marston, like other pioneers of lie detection, believed that the conscious mind could be circumvented, and the truth uncovered, through the measurement of bodily signals.

This was not a new idea. In 1730, Daniel Defoe published "An Ef-fectual Scheme for the Immediate Preventing of Street Robberies and Suppressing All Other Disorders of the Night," in which he pro-posed an alternative to physical coercion: "Guilt carries fear always about with it, there is a tremor in the blood of a thief, that, if at-tended to, would effectually discover him; and if charged as a suspi-cious fellow, on that suspicion only I would feel his pulse."

In the late nineteenth century, the Italian criminologist Cesare Lombroso invented his own version of a lie detector, based on the physiology of emotion. A suspect was told to plunge his hand into a tank filled with water, and the subject's pulse would cause the level of liquid to rise and fall slightly; the greater the fluctuation, the more dishonest the subject was judged to be.

Lombroso's student Angelo Mosso, a physiologist, noticed that shifts in emotion were often detectable in fair-skinned people in the flushing or blanching of their faces. Based on this observation, he designed a bed that rested on a fulcrum. If a suspect reclining on it told a lie, Mosso hypothesized, resulting changes in blood flow would alter the distribution of weight on the bed, unbalancing it. The device, known as Mosso's cradle, apparently never made it past the prototype.

William Moulton Marston was born in 1893, in Boston. He attended Harvard, where he worked in the lab of Hugo Münsterberg, a German émigré psychologist, who had been tinkering with an apparatus that registered responses to emotions, such as horror and tenderness, through graphical tracing of pulse rates. One student volunteer was Gertrude Stein. (She later wrote of the experience in the third person: "Strange fancies begin to crowd upon her, she feels that the silent pen is writing on and on forever.")

In 1917, Marston published a paper arguing that systolic blood pressure could be monitored to detect deception. As Ken Alder, a history professor at Northwestern, notes in his recent book, *The Lie Detectors: The History of an American Obsession*, Münsterberg and Marston's line of inquiry caught the imagination of police detectives, reporters, and law-enforcement reformers across the country, who saw a lie-detecting machine as an alternative not only to the brutal interrogation known as the third degree but also to the jury system. In 1911, an article in the *Times* predicted a future in which "there will be no jury, no horde of detectives and witnesses, no charges and countercharges, and no attorney for the defense. These impediments of our courts will be unnecessary. The State will merely submit all suspects in a case to the tests of scientific instruments."

John Larson, a police officer in Berkeley, California, who also had a doctorate in physiology, expanded on Marston's work. He built an unwieldy device, the "cardio-pneumo-psychograph," which used a standard cuff to measure blood pressure, and a rubber hose wrapped around the subject's chest to measure his breathing. Subjects were told to answer yes-or-no questions; their physiological responses were recorded by styluses that scratched black recording paper on revolving drums.

In 1921, as Alder writes, Larson had his first big chance to test his device. He was seeking to identify a thief at a residence hall for female students at Berkeley. Larson gave several suspects a six-minute exam, in which he asked various questions: "How much is thirty times forty?" "Will you graduate this year?" "Do you dance?" "Did you steal the money?" The result foretold the way in which a polygraph would often "work": as a goad to confession. A student nurse confessed to the crime—a few days after she'd stormed out during the exam.

In the early '20s, another member of the Berkeley police force, Leonarde Keeler, increased the number of physical signs that the lie detector monitored. His portable machine recorded pulse rate, blood pressure, respiration, and "electrodermal response"—again, palm sweat. Today's lie detector looks much like Keeler's eighty-year-old invention. And it bears the same name: the polygraph.

Polygraphs never caught on in Europe. But here their advent coincided with the Prohibition-era crime wave; with a new fascination with the unconscious (this was also the era of experimentation with so-called truth serums); and with the wave of technological innovation that had brought Americans electricity, radios, telephones, and cars. The lie detector quickly insinuated itself into American law enforcement: at the end of the '30s, a survey of thirteen city police departments showed that they had given polygraphs to nearly nine thousand suspects.

In 1923, Marston tried without success to get a polygraph test introduced as evidence in the Washington, D.C., murder trial of James Alphonso Frye. In its ruling, the Court of Appeals for the D.C. Circuit declared that a new scientific method had to have won "general acceptance" from experts before judges could give it credence. Since this decision, the polygraph has been kept out of most courtrooms, but there is an important exception: about half the states allow a defendant to take the test, generally on the understanding that the charges will be dropped if he passes and the results may be entered as evidence if he fails.

The polygraph became widely used in government and in business, often with dubious results. In the '50s, the State Department deployed the lie detector to help purge suspected homosexuals. As late as the '70s, a quarter of American corporations used the polygraph on their employees. Although Congress banned most such tests when it passed the Polygraph Protection Act, in 1988, the federal government still uses the polygraph for security screenings—despite high-profile mistakes. The polygraph failed to cast suspicion on Aldrich Ames, the CIA agent who spied for the Soviets, and wrongly implicated Wen Ho Lee, the Department of Energy scientist, as an agent of the Chinese government.

One excellent way to gauge the polygraph's effectiveness would be to compare it with an equally intimidating fake machine, just as a drug is compared with a placebo. But, strangely, no such experiment has ever been performed. In 1917, the year that Marston published his first paper on lie detection, his research encountered strong skepticism. John F. Shepard, a psychologist at the University of Michigan, wrote a review of Marston's research. Though the physical changes that the machine measured were "an index of activity," Shepard wrote, the same results "would be caused by so many different circumstances, anything demanding equal activity (intelligence or emotional)." The same criticism holds true today. All the physiological responses measured by the polygraph have causes other than lying, vary greatly among individuals, and can be affected by conscious effort. Breathing is particularly easy to regulate. Advice on how to beat the lie detector is a cottage industry. *Deception Detection: Winning the Polygraph Game* (1991) warns potential subjects, "Don't complain about a dry mouth. An examiner will interpret this as fear of being found out and will press you even harder." (Many people do get dry-mouthed when they're nervous—which is apparently why, during the Inquisition, a suspect was sometimes made to swallow a piece of bread and cheese: if it stuck in his throat, he was deemed guilty.)

Other well-known "countermeasures" include taking a mild sedative; using mental imagery to calm yourself; and biting your tongue to make yourself seem anxious in response to random questions.

Why, then, is the polygraph still used? Perhaps the most vexing thing about the device is that, for all its flaws, it's not pure hokum: a racing pulse and an increased heart rate can indicate guilt. Every liar has felt an involuntary flutter, at least once. Yet there are enough exceptions to insure that the polygraph will identify some innocent people as guilty and some guilty people as innocent.

At the Cambridge conference, Jed S. Rakoff, a United States district judge in New York, told a story about a polygraph and a false confession. Days after September 11th, an Egyptian graduate student named Abdallah Higazy came to the attention of the FBI. Higazy had been staying at the Millennium Hotel near Ground Zero on the day of the attacks. A hotel security guard claimed that he had found a pilot's radio in Higazy's room. Higazy said that it wasn't his, and when he appeared before Rakoff he asked to be given a polygraph. As Rakoff recalled, "Higazy very much believed in them and thought it would exonerate him." During a four-hour interrogation by an FBI polygrapher, Higazy first repeated that he knew nothing about the radio, and then said that maybe it was his. He was charged with lying to the FBI and went to prison. Within a month, a pilot stopped by the hotel to ask about a radio that he had accidentally left there. The security guard who found the radio admitted that it hadn't been in Higazy's room; he was prosecuted and pled guilty. Higazy was exonerated, and a subsequent investigation revealed that he had felt dizzy and ill during the examination, probably out of nervousness. But when Higazy asked the polygrapher if anyone had ever become ill during a polygraph test he was told that "it had not happened to anyone who told the truth."

TO DATE, THERE HAVE BEEN only a dozen or so peer-reviewed studies that attempt to catch lies with fMRI technology, and most of them involved fewer than twenty people. Nevertheless, the idea has inspired a torrent of media attention, because scientific studies involving brain scans dazzle people, and because mind reading by machine is a beloved science-fiction trope, revived most recently in movies like *Minority Report* and *Eternal Sunshine of the Spotless Mind*. Many journalistic accounts of the new technology—accompanied by colorful bitmapped images of the brain in action—resemble science fiction themselves. In January, the *Financial Times* proclaimed, "For the first time in history, it is becoming possible to read someone else's mind with the power of science." A CNBC report, accompanied by the Eurythmics song "Would I Lie to You?" showed its reporter entering an fMRI machine, described as a "sure-fire way to identify a liar." In March, a cover story in the *New York Times Magazine* predicted transformations of the legal system in response to brain imaging; its author, Jeffrey Rosen, suggested that there was a widespread "fear" among legal scholars that "the use of brain-scanning technology as a kind of super mind-reading device will threaten our privacy and mental freedom." *Philadelphia* has declared "the end of the lie," and a *Wired* article, titled "Don't Even Think About Lying," proclaimed that fMRI is "poised to transform the security industry, the judicial system, and our fundamental notions of privacy." Such talk has made brain-scan lie detection sound as solid as DNA evidence—which it most definitely is not.

Paul Bloom, a cognitive psychologist at Yale, believes that brain imaging has a beguiling appeal beyond its actual power to explain mental and emotional states. "Psychologists can be heard grousing that the only way to publish in *Science* or *Nature* is with pretty color pictures of the brain," he wrote in an essay for the magazine *Seed*. "Critical funding decisions, precious column inches, tenure posts, science credibility, and the popular imagination have all been influenced by fMRI's seductive but deceptive grasp on our attentions." Indeed, in the past decade, *Nature* alone has published nearly a hundred articles involving fMRI scans. The technology is a remarkable tool for exploring the brain, and may one day help scientists understand much more about cognition and emotion. But enthusiasm for brain scans leads people to overestimate the accuracy with which they can pinpoint the sources of complex things like love or altruism, let alone explain them.

Brain scans enthrall us, in part, because they seem more like "real" science than those elaborate deductive experiments that so many psychologists perform. In the same way that an X-ray confirms a bone fissure, a brain scan seems to offer an objective measure of mental activity. And, as Bloom writes, fMRI research "has all the trappings of work with great lab-cred: big, expensive, and potentially dangerous machines, hospitals and medical centers, and a lot of people in white coats."

Deena Skolnick Weisberg, a graduate student at Yale, has conducted a clever study, to be published in the *Journal of Cognitive Neuroscience*, which points to the outsized glamour of brain-scan research. She and her colleagues provided three groups—neuroscientists, neuroscience students, and ordinary adults—with explanations for common psychological phenomena (such as the tendency to assume that other people know the same things we do). Some of these explanations were crafted to be bad. Weisberg found that all three groups were adept at identifying the bad explanations, except when she inserted the words "Brain scans indicate." Then the students and the regular adults became notably less discerning. Weisberg and her colleagues conclude, "People seem all too ready to accept explanations that allude to neuroscience."

Some bioethicists have been particularly credulous, assuming that MRI mind reading is virtually a done deal, and arguing that there is a need for a whole new field: "neuroethics." Judy Illes and Eric Racine, bioethicists at Stanford, write that fMRI, by laying bare the brain's secrets, may "fundamentally alter the dynamics between personal identity, responsibility, and free will." A recent article in the *American Journal of Bioethics* asserts that brain-scan lie detection may "force a reexamination of the very idea of privacy, which up until now could not reliably penetrate the individual's cranium."

Legal scholars, for their part, have started debating the constitutionality of using brain-imaging evidence in court. At a recent meeting of a National Academy of Sciences committee on lie detection, in Washington, D.C., Hank Greely, a Stanford law professor, said, "When we make speculative leaps like these . . . it increases, sometimes in detrimental ways, the belief that the technology works." In the rush of companies like No Lie to market brain scanning, and in the rush of scholars to judge the propriety of using the technology, relatively few people have asked whether fMRIs can actually do what they either hope or fear they can do.

Functional MRI is not the first digital-age breakthrough that was supposed to supersede the polygraph. First, there was "brain fingerprinting," which is based on the idea that the brain releases a recognizable electric signal when processing a memory. The technique used EEG sensors to try to determine whether a suspect retained memories related to a crime—an image of, say, a murder weapon. In 2001, *Time* named Lawrence Farwell, the developer of brain fingerprinting, one of a hundred innovators who "may be the Picassos or the Einsteins of the 21st century." But researchers have since noted a big drawback: it's impossible to distinguish between brain signals produced by actual memories and those produced by imagined memories—as in a made-up alibi.

After September 11th, another technology was widely touted: thermal imaging, an approach based on the finding that the area around the eyes can heat up when people lie. The developers of this method—Ioannis Pavlidis, James Levine, and Norman Eberhardt—published journal articles that had titles like "Seeing Through the Face of Deception" and were accompanied by dramatic thermal images. But the increased blood flow that raises the temperature around the eyes is just another mark of stress. Any law-enforcement agency that used the technique to spot potential terrorists would also pick up a lot of jangly, harmless travelers.

Daniel Langleben, the Penn psychiatrist whose research underpins No Lie, began exploring this potential new use for MRIs in the late

'90s. Langleben, who is forty-five, has spent most of his career studying the brains of heroin addicts and hyperactive boys. He developed a side interest in lying partly because his research agenda made him think about impulse control, and partly because his patients often lied to him. Five years ago, Langleben and a group of Penn colleagues published the study on brain scanning and lie detection that attracted Huizenga's attention. In the experiment, which was written up in *NeuroImage*, each of twenty-three subjects was offered an envelope containing a twenty-dollar bill and a playing card—the five of clubs. They were told that they could keep the money if they could conceal the card's identity when they were asked about it inside an MRI machine. The subjects pushed a button to indicate yes or no as images of playing cards flashed on a screen in front of them. After Langleben assembled the data, he concluded that lying seemed to involve more cognitive effort than truthtelling, and that three areas of the brain generally became more active during acts of deception: the anterior cingulate cortex, which is associated with heightened attention and error monitoring; the dorsal lateral prefontal cortex, which is involved in behavioral control; and the parietal cortex, which helps process sensory input. Three years later, Langleben and his colleagues published another study, again involving concealed playing cards, which suggested that lying could be differentiated from truthtelling in individuals as well as in groups. The fMRI's accuracy rate for distinguishing truth from lies was 77 percent.

Andrew Kozel and Mark George, then at the Medical University of South Carolina, were doing similar work at the time; in 2005, they published a study of fMRI lie detection in which thirty people were instructed to enter a room and take either a watch or a ring that had been placed there. Then, inside a scanner, they were asked to lie about which object they had taken but to answer truthfully to neutral questions, such as "Do you like chocolate?" The researchers distinguished truthful from deceptive responses in 90 percent of the cases. (Curiously, Kozel's team found that liars had heightened activity in different areas of the brain than Langleben did.)

Langleben and Kozel weren't capturing a single, crisp image of the brain processing a lie; an fMRI's record of a split-second event is considered unreliable. Instead, they asked a subject to repeat his answer dozens of times while the researchers took brain scans every couple of seconds. A computer then counted the number of "voxels" (the 3-D version of pixels) in the brain image that reflected a relatively high level of oxygenated blood, and used algorithms to determine whether this elevated activity mapped onto specific regions of the brain.

One problem of fMRI lie detection is that the machines, which cost about three million dollars each, are notoriously finicky. Technicians say that the scanners often have "bad days," in which they can produce garbage data. And a subject who squirms too much in the scanner can invalidate the results. (Even moving your tongue in your mouth can cause a problem.) The results for four of the twenty-three subjects in Langleben's first study had to be thrown out because the subjects had fidgeted.

The Langleben studies also had a major flaw in their design: the concealed playing card came up only occasionally on the screen, so the increased brain activity that the scans showed could have been a result not of deception but of heightened attention to the salient card. Imagine that you're the research subject: You're lying on your back, trying to hold still, probably bored, maybe half asleep, looking at hundreds of cards that don't concern you. Then, at last, up pops the five of clubs—and your brain sparks with recognition.

Nearly all the volunteers for Langleben's studies were Penn students or members of the academic community. There were no sociopaths or psychopaths; no one on antidepressants or other psychiatric medication; no one addicted to alcohol or drugs; no one with a criminal record; no one mentally retarded. These allegedly seminal studies look exclusively at unproblematic, intelligent people who were instructed to lie about trivial matters in which they had little stake. An incentive of twenty dollars can hardly be compared with, say, your freedom, reputation, children, or marriage—any or all of which might be at risk in an actual lie-detection scenario.

The word "lie" is so broad that it's hard to imagine that any test, even one that probes the brain, could detect all forms of deceit: small, polite lies; big, brazen, self-aggrandizing lies; lies to protect or enchant our children; lies that we don't really acknowledge to ourselves as lies; complicated alibis that we spend days rehearsing. Certainly, it's hard to imagine that all these lies will bear the identical neural signature. In their degrees of sophistication and detail, their moral weight, their emotional valence, lies are as varied as the people who tell them. As Montaigne wrote, "The reverse side of the truth has a hundred thousand shapes and no defined limits."

Langleben acknowledges that his research is not quite the breakthrough that the media hype has suggested. "There are many questions that need to be looked into before we know whether this will work as lie detection," he told me. "Can you do this with somebody who has an IQ of ninety-five? Can you do it with somebody who's fifty or older? Somebody who's brain-injured? What kinds of real crimes could you ask about? What about countermeasures? What about people with delusions?"

Nevertheless, the University of Pennsylvania licensed the pending patents on his research to No Lie in 2003, in exchange for an equity position in the company. Langleben didn't protest. As he explained to me, "It's good for your résumé. We're encouraged to have, as part of our portfolio, industry collaborations." He went on, "I was trying to be a good boy. I had an idea. I went to the Center of Technology Transfer and asked them, 'Do you like this?' They said, '*Yeah*, we like that.'"

STEVEN LAKEN IS THE CEO of Cephos, a Boston-based company that is developing a lie-detection product based on Kozel's watch-and-ring study. (It has an exclusive licensing agreement for pending patents that the Medical University of South Carolina applied for in 2002.) Cephos is proceeding more cautiously than No Lie. Laken's company is still conducting studies with Kozel, the latest of which involve more than a hundred people. (The sample pool is again young, healthy, and free of criminal records and psychological problems.) Cephos won't be offering fMRIs commercially until the results of those studies are in; Laken predicts that this will happen within a year. At the National Academy of Sciences committee meeting, he said, "I can say we're not at 90 percent accuracy. And I have said, if we were not going to get to 90 percent, we're not going to sell this product." (Nobody involved in fMRI lie detection seems troubled by a 10 percent error rate.)

In March, I went to a suburb of Boston to meet Laken. He is thirty-five years old and has a Ph.D. in cellular and molecular medicine from Johns Hopkins. Nine years ago, he identified a genetic mutation that can lead to colorectal cancer. He has a more conservative temperament than Joel Huizenga does, and he told me he thinks that spousal-fidelity cases are "sleazy." But he sees a huge potential market for what he calls a "truth verifier"—a service for people looking to exonerate themselves. "There are some thirty-five million criminal and civil cases filed in the U.S. every year," Laken said. "About twenty million are criminal cases. So let's just say that you never even do a criminal case—well, that still leaves roughly fifteen million for us to go after. Some you exclude, but you end up with several million cases that are high stakes: two people arguing about things that are important." Laken also thinks that fMRI lie detection could help the government elicit information, and confessions, from terrorist suspects, without physical coercion.

He calmly dismissed the suggestion that the application of fMRI lie detection is premature. "I've heard it said, 'This technology can't work because it hasn't been tested on psychopaths, and it hasn't been tested on children, and it certainly hasn't been tested on psychopathic children,'" he said. "If that were the standard, there'd never be any medicine."

Laken and I spoke while driving to Framingham, Massachusetts, to visit an MRI testing center run by Shields, a company that operates twenty-two such facilities in the state. Laken was working on a deal with Shields to use their scanners. For Shields, it would be a smart move, Laken said, because customers would pay up front for the scan—there would be no insurance companies to contend with. (Cephos and Shields have since made an official arrangement.) Laken believes that Cephos will prosper primarily through referrals: lawyers will function as middlemen, ordering an fMRI for a client, much as a doctor orders an MRI for a patient.

We pulled into the parking lot, where a sign identifies Shields as "the MRI provider for the 3-X World Champion New England Patriots." Inside, John Cannillo, an imaging specialist at Shields, led us into a room to observe a woman undergoing an MRI exam. She lay on a platform that slid into a white tubular scanner, which hummed like a giant tuning fork.

During a brain scan, the patient wears a copper head coil, in order to enhance the magnetic field around the skull. The magnet is so powerful that you have to remove any metal objects, or you will feel a tugging sensation. If a person has metal in his body—for instance, shrapnel, or the gold grillwork that some hip-hop fans have bonded to their teeth—it can pose a danger or invalidate the results. At the NAS meeting in Washington, one scientist wryly commented, "It could become a whole new industry—criminals having implants put in to avoid scanning."

A Shields technician instructed the woman in the scanner from the other side of a glass divide. "Take a breath in, and hold it, hold it," he said. Such exercises help minimize a patient's movements.

As we watched, Laken admitted that "the kinks" haven't been worked out of fMRI lie detection. "We make mistakes," he said of his company. "We don't know why we make mistakes. We may *never* know why. We hope we can get better." Some bioethicists and journalists may worry about the far-off threat to "cognitive freedom," but the real threat is simpler and more immediate: the commercial introduction of high-tech "truth verifiers" that may work no better than polygraphs but seem more impressive and scientific. Polygraphs, after all, are not administered by licensed medical professionals.

Nancy Kanwisher, a cognitive scientist at MIT, relies a great deal on MRI technology. In 1997, she identified an area near the bottom of the brain that is specifically involved in perceiving faces. She has become a pointed critic of the rush to commercialize brain imaging for lie detection, and believes that it's an exaggeration even to say that research into the subject is "preliminary." The tests that have been done, she argues, don't really look at lying. "Making a false response when instructed to do so is not a *lie*," she says. The 90 percent "accuracy" ascribed to fMRI lie detection refers to a scenario so artificial that it is nearly meaningless. To know whether the technology works, she believes, "you'd have to test it on people whose guilt or innocence hasn't yet been determined, who believe the scan will reveal their guilt or innocence, and whose guilt or innocence can be established by other means afterward." In other words, you'd have to run a legal version of a clinical trial, using real suspects instead of volunteers.

Langleben believes that Kanwisher is too pessimistic. He suggested that researchers could recruit people who had been convicted of a crime in the past and get them to lie retrospectively about it. Or maybe test subjects could steal a "bagel or something" from a convenience store (the researchers could work out an agreement with the store in advance) and then lie about it. But even these studies don't approximate the real-world scenarios Kanwisher is talking about.

She points out that the various brain regions that appear to be significantly active during lying are "famous for being activated in a wide range of different conditions—for almost any cognitive task that is more difficult than an easier task." She therefore believes that fMRI lie detection would be vulnerable to countermeasures—performing arithmetic in your head, reciting poetry—that involve concerted cognitive effort. Moreover, the regions that allegedly make up the brain's "lying module" aren't that small. Even Laken admitted as much. As he put it, "Saying 'You have activation in the anterior cingulate' is like saying 'You have activation in Massachusetts.'"

Kanwisher's complaint suggests that fMRI technology, when used cavalierly, harks back to two pseudosciences of the eighteenth and nineteenth centuries: physiognomy and phrenology. Physiognomy held that a person's character was manifest in his facial features; phrenology held that truth lay in the bumps on one's skull. In 1807, Hegel published a critique of physiognomy and phrenology in *The Phenomenology of Spirit*. In that work, as the philosopher Alasdair MacIntyre writes, Hegel observes that "the rules that we use in everyday life in interpreting facial expression are highly fallible." (A friend who frowns throughout your piano recital might explain that he was actually fuming over an argument with his wife.) Much of what Hegel had to say about physiognomy applies to modern attempts at mind reading. Hegel quotes the scientist Georg Christoph Lichtenberg, who, in characterizing physiognomy, remarked, "If anyone said, 'You act, certainly, like an honest man, but I can see from your face you are forcing yourself to do so, and are a rogue at heart,' without a doubt every brave fellow to the end of time when accosted in that fashion will retort with a box on the ear." This response is correct, Hegel argues, because it "refutes the fundamental assumption of such a 'science' of conjecture—that the reality of a man is his face, etc. The true being of man is, on the contrary, his act; individuality is real in the deed." In a similar vein, one might question the core presumption of fMRI—that the reality of man is his brain.

Elizabeth Phelps, a prominent cognitive neuroscientist at NYU, who studies emotion and the brain, questions another basic assumption behind all lie-detection schemes—that telling a falsehood creates conflict within the liar. With the polygraph, the assumption is that the conflict is emotional: the liar feels guilty or anxious, and these feelings produce a measurable physiological response. With brain imaging, the assumption is that the conflict is cognitive: the liar has to work a little harder to make up a story, or even to stop himself from telling the truth. Neither is necessarily right. "Sociopaths don't feel the same conflict when they lie," Phelps says. "The regions of the brain that might be involved if you have to inhibit a response may not be the same when you're a sociopath, or autistic, or maybe just strange. Whether it's an emotional or a cognitive conflict you're supposed to be exhibiting, there's no reason to assume that your response wouldn't vary depending on what your personal tendencies are—on who *you* are."

WHEN I TALKED TO HUIZENGA, the No Lie CEO, a few months after I had met him in California, he was unperturbed about the skepticism that he was encountering from psychologists. "In science, when you go out a little further than other people, it can be hard," he said. "The top people understand, but the middle layer don't know what you're talking about."

Huizenga told me that he was trying to get fMRI evidence admitted into a California court for a capital case that he was working on. (He would not go into the case's details.) Given courts' skepticism toward the polygraph, Huizenga's success is far from certain. Then again we are in a technology-besotted age that rivals the '20s, when Marston popularized lie detection. And we live in a time when there is an understandable hunger for effective ways to expose evildoers, and when concerns about privacy have been nudged aside by our desire for security and certainty. "Brain scans indicate": what a powerful phrase. One can easily imagine judges being impressed by these pixellated images, which appear so often in scientific journals and in the newspaper. Indeed, if fMRI lie detection is successfully marketed as a service that lawyers steer their clients to, then a refusal even to take such a test could one day be cause for suspicion.

Steven Hyman, the Harvard psychiatrist, is surprised that companies like No Lie have eluded government oversight. "Think of a medical test," he said. "Before it would be approved for wide use, it would have to be shown to have acceptable accuracy among the populations in whom it would be deployed. The published data on the use of fMRI for lie detection uses highly artificial tests, which are not even convincing models of lying, in very structured laboratory settings. There are no convincing data that they could be used accurately to screen a person in the real world." But, in the end, that might not matter. "Pseudo-colored pictures of a person's brain lighting up are undoubtedly more persuasive than a pattern of squiggles produced by a polygraph," he said. "That could be a big problem if the goal is to get to the truth."

Laken, meanwhile, thinks that people who find themselves in a jam, and who are desperate to exonerate themselves, simply have to educate themselves as consumers. "People have said that fMRI tests are unethical and immoral," he said. "And the question is, Why is it unethical and immoral if somebody wants to spend their money on a test, as long as they understand what it is they're getting into? We've never said the test was *perfect*. We've never said we can guarantee that this is admissible in court and that's it—you're scot-free." Later that day, I looked again at the Cephos Web site. It contained a bolder proclamation. "The objective measure of truth and deception that Cephos offers," it said, "will help protect the innocent and convict the guilty."

STEPHEN S. HALL

The Older-and-Wiser Hypothesis

FROM THE *NEW YORK TIMES MAGAZINE*

It seems these days that no matter how "fuzzy" or abstract a con-cept, there are scientists seeking to measure it. So, too, with the nature of "wisdom," and the question of whether it increases with age, which researchers are now studying. As Stephen S. Hall discov-ers, even if the results of this investigation are not conclusive, they still throw off some meaningful surprises.

IN 1950, THE PSYCHOANALYST Erik H. Erikson, in a famous treatise on the phases of life development, identified wisdom as a likely, but not inevitable, by-product of growing older. Wisdom arose, he suggested, during the eighth and final stage of psychosocial development, which he described as "ego integrity versus despair." If an individual had achieved enough "ego integrity" over the course of a lifetime, then the imminent approach of infirmity and death would be accompanied by the virtue of wisdom. Unfortunately for researchers who followed, Erikson didn't bother to define wisdom.

As an ancient concept and esteemed human value, wisdom has historically been studied in the realms of philosophy and religion. The idea has been around at least since the Sumerians first etched bits of practical advice—"We are doomed to die; let us spend"—on clay tablets more than five thousand years ago. But as a trait that might be captured by quantitative measures, it has been more like the woolly mammoth of ideas—big, shaggy, and elusive. It is only in the last three decades that wisdom has received even glancing attention from social scientists. Erikson's observations left the door open for the formal study of wisdom, and a few brave psychologists rushed in where others feared to tread.

In some respects, they have not moved far beyond the very first question about wisdom: What is it? And it won't give anything away to reveal that thirty years after embarking on the empirical study of wisdom, psychologists still don't agree on an answer. But it is also true that the journey in many ways may be as enlightening as the destination.

From the outset, it's easier to define what wisdom isn't. First of all, it isn't necessarily or intrinsically a product of old age, although reaching an advanced age increases the odds of acquiring the kinds of life experiences and emotional maturity that cultivate wisdom, which is why aspects of wisdom are increasingly attracting the attention of gerontological psychologists. Second, if you think you're wise, you're probably not. As Gandhi (who topped the leader board a few years ago in a survey in which college students were asked to name wise people) put it, "It is unwise to be too sure of one's own wisdom." Indeed, a general thread running through modern wisdom research is that wise people tend to be humble and "other-centered" as opposed to self-centered.

"Wisdom is really hard to study—really hard," says Robert J. Sternberg, a former president of the American Psychological Association who edited *Wisdom: Its Nature, Origins, and Development,* one of the first academic books on the subject, in 1990, and also edited, with Jennifer Jordan, *A Handbook of Wisdom* in 2005. "People tend to pooh-pooh wisdom because, well, you know, what's that? And how could you possibly define it? Isn't it culturally relative?" And yet Sternberg, who is the dean of the School of Arts and Sciences at Tufts University, says he believes the cultivation of wisdom—even though the concept is "big, important, and messy"—is essential to the future of society.

Certain qualities associated with wisdom recur in the academic literature: a clear-eyed view of human nature and the human predicament; emotional resiliency and the ability to cope in the face of adversity; an openness to other possibilities; forgiveness; humility; and a knack for learning from lifetime experiences. And yet as psychologists have noted, there is a yin-yang to the idea that makes it difficult to pin down. Wisdom is founded upon knowledge, but part of the physics of wisdom is shaped by uncertainty. Action is important, but so is judicious inaction. Emotion is central to wisdom, yet detachment is essential.

If you think all those attributes sound fuzzy, vague, and absolutely refractory to quantification, you've got a lot of company in the academic community. But there is a delicious paradox at the heart of the study of wisdom. As difficult as it is to define, the mere contemplation of a definition is an irresistible exercise that says a lot about who we aspire to become over the course of a lifetime and what we value as a society. And little pieces of that evolving definition of wisdom—especially the ability to cope with adversity and the regulation of emotion with age—have begun to attract researchers with brain-scanning machines and serious chops in neuroscience.

"It's very intriguing, and it's becoming a big issue in our field," says Suzanne Kunkel, director of the Scripps Gerontology Center at Miami University in Ohio. She noted that the number of formal talks about wisdom and the aging process has increased significantly at professional meetings. "Part of me is a little skeptical," she says, reflecting the compelling ambivalence the subject elicits, "and part of me thinks there's something there."

THE FORMAL STUDY OF wisdom as a modern academic pursuit can legitimately trace its roots back to the 1950s, to an apartment building on Newkirk Avenue, just off Coney Island Avenue in Brooklyn. That is where a keenly observant young girl named Vivian Clayton became fascinated by special qualities she attributed to two prominent elders in her life: her father, a furrier named Simon Clayton, and her maternal grandmother. There was something that distinguished them from everyone else she knew. Despite limited education, they possessed an uncanny ability to remain calm in the midst of crises, made good decisions, and conveyed an almost palpable sense of emotional contentment, often in the face of considerable adversity or uncertainty. Long before she went to college, Clayton found herself contemplating the nature of wisdom.

"My father was forty-one when I was born," she said recently. "By far, he was the oldest parent among all my friends, almost the age of my friends' grandparents. He had emigrated from England but had lived through World War II there and experienced the blitz and had to care for his dying mother, who was so sick that she refused to go down into the shelters during air raids in London. She lived in the East End, where the docks were, and they were always getting bombed. So he would sit with her while the bombs were falling, and when it was over, she would say, 'Now we can have a cup of tea!' He was a very humble man, and very aware of his limitations, but he always seemed to be able to weigh things and then make decisions that were right for the family. He knew what to respond to quickly, and what you had to reflect on." Clayton's maternal grandmother, Beatrice Domb, was the other central figure in her early life. "My mother saw my grandmother as a simple person," Clayton says. "But her simplicity I saw as a sign of deep contentment in her own life. She, who had less than a high-school education, was the matriarch of this very large family."

During her childhood and adolescence, Clayton obsessed over the differences between her mother and father, her grandmother and grandfather. She recalls pondering these differences as a teenager, dipping her toes in Mahwah Creek during family outings in Suffern, northwest of the city; as an undergraduate studying psychology at Buffalo University; and more formally, as a graduate student in the early 1970s at the University of Southern California, working with one of the country's leading gerontological psychologists, James E. Birren. Clayton is generally recognized as the first psychologist to ask, in even faintly scientific terms, "What does wisdom mean, and how does age affect it?"

Clayton's study of wisdom began with a bias, but one that counterbalanced a preexisting bias that pervaded the biomedical literature on aging in the '60s and '70s. Half a century ago, although only 5 percent of the elderly lived in nursing homes, almost all the gerontological research focused on this frail and struggling population. Not surprisingly, these researchers found plenty of negative things about being old. Memory, especially working memory, began to fade. The speed with which the brain processed information slowed down. Older people were more likely to be cognitively impaired.

One of the leading voices pushing for a more balanced view of the aging process was Birren. In what might be viewed as a battle between modern psychology and cultural attitudes toward the elderly, Birren was one of the leaders of an effort to investigate positive aspects of aging. At the time Clayton was at USC, Birren's graduate students were exploring the relationship of aging to topics like love, creativity, and wisdom—topics so big and unwieldy that they almost defied study.

Clayton went off to consult the "literature" on wisdom, which almost mirrored the central canon of Western civilization. She rummaged through the Hebrew Bible for clues to wise behavior, analyzed the stories of Job and King Solomon, parsed the meaning of ancient proverbs. "What emerged from that analysis," she says, "was that wisdom meant a lot of different things. But it was always associated with knowledge, frequently applied to human social situations, involved judgment and reflection, and was almost always embedded in a component of compassion." The essential importance of balance was embodied in the Hebrew word for wisdom, *chochmah*, which ancient people understood to evoke the combination of both heart and mind in reaching a decision. At that point, Birren advised Clayton to "become more scientific" and treat wisdom as a psychological construct that could be defined well enough to be measured and studied ("operationalized," in psychological lingo).

Between 1976, when she finished her dissertation, and 1982, Clayton published several groundbreaking papers that are now generally acknowledged as the first to suggest that researchers could study wisdom empirically. She identified three general aspects of human activity that were central to wisdom—the acquisition of knowledge (cognitive) and the analysis of that information (reflective) filtered through the emotions (affective). Then she assembled a battery of existing psychological tests to measure it.

Clayton laid several important markers on the field at its inception. She realized that "neither were the old always wise, nor the young lacking in wisdom." She also argued that while intelligence represented a nonsocial and impersonal domain of knowledge that might diminish in value over the course of a lifetime, wisdom represented a social, interpersonal form of knowledge about human nature that resisted erosion and might increase with age. Clayton's early work was "a big deal," Sternberg says. "It was a breakthrough to say wisdom is something you could study." Jacqui Smith, who has conducted wisdom research since the 1980s, says it "was seminal work that really triggered subsequent studies."

As Clayton began to describe her research at psychological meetings in the late '70s, the work on wisdom created considerable buzz. One of the people who grasped its significance immediately was Paul B. Baltes, a legendary psychologist then at Pennsylvania State

University. Baltes helped pioneer life-span developmental theory, which argues that in order to understand, say, a sixty-year-old person, you need to take into account the individual's biology, psychology, and sociological context at various stages of life, as well as the cultural and historical era in which he or she lived.

Baltes closely monitored the initial wisdom studies, Clayton recalls, and regularly peppered her with questions about her progress. "I went to all these meetings," she says, "and we would have lunch or dinner at every meeting. He was always asking, where was I with this wisdom stuff?"

The answer would soon be: nowhere. In 1982, Clayton published her last paper on wisdom. By then, she had applied for, but failed to receive, a grant from the National Institute on Aging to pursue the wisdom studies, had quit her position as assistant professor at Columbia University Teachers College and left academia for good. Part of the reason was that she recognized her own limitations in studying a very diffuse topic. "I was lost in the Milky Way of wisdom," she admits, "and each star seemed as bright as the next. Ultimately that's why I didn't continue with it." The universe shifted to Berlin, and the working definition of wisdom acquired a German accent.

THE BERLIN WISDOM PARADIGM, as it came to be called, was built in part on research using hypothetical vignettes to discern wise and unwise responses to life dilemmas. "A fifteen-year-old girl wants to get married right away," one vignette suggested. "What should one/she consider and do?"

A wise person, according to the Berlin group, would say something like: "Well, on the surface, this seems like an easy problem. On average, marriage for fifteen-year-old girls is not a good thing. But there are situations where the average case does not fit. Perhaps in this instance, special life circumstances are involved, such as the girl has a terminal illness. Or the girl has just lost her parents. And also this girl may live in another culture or historical period. Perhaps she was raised with a value system different from ours. In addition, one has to think about adequate ways of talking with the girl and to consider her emotional state."

That reply may seem tentative and relativistic, but it reflects many aspects of wisdom as defined by the Berlin Wisdom Project, which began in 1984 under the leadership of Baltes, who along with Birren had championed the search for late-life potential. Born in 1939 in Germany, Baltes had established a reputation as a leading quantitative psychologist by the time he returned to Germany in 1980 to become director of the Max Planck Institute for Human Development in Berlin. There, Baltes and many collaborators—including Jacqui Smith (now at the University of Michigan), Ursula M. Staudinger, and Ute Kunzmann—embarked on an ambitious, large-scale program to, as they put it, "take wisdom into the laboratory."

Boiled down to its essence, the "Berlin Paradigm" defined wisdom as "an expert knowledge system concerning the fundamental pragmatics of life." Heavily influenced by life-span psychology, the Berlin version of wisdom emphasized several complementary qualities: expert knowledge of both the "facts" of human nature and the "how" of dealing with decisions and dilemmas; an appreciation of one's historical, cultural, and biological circumstances during the arc of a life span; an understanding of the "relativism" of values and priorities; and an acknowledgment, at the level of both thought and action, of uncertainty. "We picked up from the philosophical literature that wisdom is like a peak performance," Smith says. "It's the highest level of potential or achievement that a human mind might be able to achieve." And so the Berlin group focused more on expertise and performance than on personality traits, because such an approach lent itself to more rigorous measurement than the typical self-report tests of psychological research.

"Wisdom in action," as the Berlin group put it, might manifest itself as good judgment, shrewd advice, psychological insight, emotional regulation and empathetic understanding; it could be found in familial interactions, in formal writing and in the relationship between a student and mentor or a doctor and patient. Yet by its very nature, the researchers conceded, wisdom was a utopian concept that was virtually unattainable. Baltes and Staudinger pointed out in one paper that "wisdom is a collectively anchored product and that individuals by themselves are only 'weak' carriers of wisdom." They generally did not see wisdom as the function of personality. As Smith puts it: "We went in the other direction and tried to define what a product might be. Not the person as such, but rather some sort of performance that we could assess." In evaluating the wisdom of Gandhi, for example, they focused on his speeches and writings.

One instrument the Baltes group developed to measure wisdom was posing open-ended, hypothetical questions like the one about the fifteen-year-old girl who wanted to marry. (In their view, a reply garnering a low wisdom-related score would be an inflexible, authoritative response like: "No, no way, marrying at age fifteen would be utterly wrong. One has to tell the girl that marriage is not possible. . . . No, this is just a crazy idea.") These vignettes located wisdom firmly in the universe of problem-solving around significant life events—from issues like choosing a career versus child-rearing to facing decisions about early retirement to dealing with a diagnosis of cancer.

The Germans were among the first to reach what is now a widespread conclusion: there's not a lot of wisdom around. Of the seven hundred people assessed, "we never found a single person who gained top scores across the board," Smith wrote in an e-mail message. They also punctured one conceit about growing old when they found no evidence, in four different studies, that wisdom, as they defined it, necessarily increases with age. Rather, they identified a "plateau" of wisdom-related performance through much of middle and old age; a separate study by the group has indicated that wisdom begins, on average, to diminish around age seventy-five, probably hand in hand with cognitive decline. Nonetheless, the Baltes group suggested in one paper that there might be an optimal age and that "the 'world record' in wisdom may be held by someone in his or her 60s."

The Berlin Wisdom Project made a huge impact on the handful of people interested in wisdom research; by one account, academic "wisdom" publications numbered only two or three a year before 1984 but had grown to several dozen a year by 2000. But the German research, though much admired, did not overcome many of the mainstream reservations in academia. Jacqui Smith, who was collaborating with Baltes on one of his final wisdom papers when he died of cancer last fall at age sixty-seven, says wisdom studies remain on the fringe of academic respectability.

Even some wisdom researchers found the Berlin wisdom studies to be abstract and difficult to understand, and although emotion was always part of the formula, it struck some people as secondary to the emphasis on expert knowledge. "It's great work, and they've looked at it more closely than anybody else," says Laura L. Carstensen, a psychologist who directs the Center on Longevity at Stanford University. "But one of the critiques people have had is that they left emotion out of it. I don't think you can have wisdom without having emotional regulation be a part of it."

HOW MIGHT EMOTION BE important to wisdom? Consider C., a sixty-seven-year-old mother of seven children who lives in Gainesville, Florida. Her life has not been without heartache or emotional tumult. She grew up poor, and she has been drawn into custody battles and financial imbroglios with in-laws. More significant, one of her children was born with cerebral palsy; rather than place the child in a home, as some urged her to do, she insisted on caring for and raising him at home with the rest of the family. "I would put my healthy kids in a home first," she told doctors at the time, "instead of putting a baby in there that can't talk for himself." Despite years of challenge (the son eventually died at age twelve), C. managed to maintain a kind of emotional equilibrium. "I don't sit around and dwell on bad things," she said. "I don't have time for it, really. There's so many good things you can do."

C., who appears as a pseudonym in the psychological literature, is arguably one of the few certifiably wise people in the world—"certified" in the sense that she scored well above average in a "Three-Dimensional Wisdom Scale" developed by Monika Ardelt, a German-born sociologist at the University of Florida in Gainesville.

In 1990, as a graduate student at the University of North Carolina, Ardelt wanted to identify factors that contributed to a sense of life satisfaction and well-being in old age and began to focus on the acquisition of wisdom. She discovered Vivian Clayton's early research, which made emotion a central part of wisdom, and she began to build upon Clayton's framework. By 1997, Ardelt had joined the faculty at the University of Florida, and she received a grant from the National Institutes of Health and the National Institute on Aging to develop a psychological test to assess wisdom. She was interested in investigating measures of wisdom and looking at a trait that often goes by the name "resilience"—how some older people are able to deal with adversity and bounce back emotionally while others cannot. Indeed, as she has noted, "successfully coping with crises and hardships in life might not only be a hallmark of wise individuals but also one of the pathways to wisdom."

Thus, beginning in December 1997, Ardelt began to recruit 180 senior citizens at churches and community groups in north-central Florida to participate in what she called a "Personality and Aging Well Study." The participants did not know that one purpose of the study was to road-test a series of questions designed to assess general wisdom. In Ardelt's working definition, wisdom integrated three separate but interconnected ways of dealing with the world: cognitive, reflective, and emotional. Hence, a "three-dimensional" wisdom scale, which, according to the habit of psychological measures, is designated "3D-WS." The cognitive aspect, for example, included the ability to understand human nature, perceive a situation clearly, and make decisions despite ambiguity and uncertainty. The reflective sphere dealt with a person's ability to examine an event from multiple perspectives—to step outside oneself and understand another point of view. And the emotional aspect primarily involved feeling compassion toward others as well as an ability to remain positive in the face of adversity. In the initial phase, participants responded to 132 questions that probed for these qualities. Later, Ardelt settled on 39 questions that, in her judgment, captured the elusive concept of wisdom.

There is, of course, something utterly quixotic about assessing human wisdom on the basis of a self-report test in which subjects agree or disagree with statements like "People are either good or bad" and "I always try to look at all sides of a problem." Yet the Three-Dimensional Wisdom Scale, Ardelt argues, distinguished "how relatively wise older people cope with life crises in comparison to older people relatively low on wisdom." And when Ardelt went back and intensively interviewed some of the subjects (including C.), a seasoned, pragmatic, everyday version of wisdom—wisdom with a small "w," you might say—emerged in their life stories.

J., who was also described in the literature, is an eighty-six-year-old African American man who is also no stranger to adversity. He went off to fight in World War II and, after experiencing the horrors of battle, suffered severe depression upon his return to the United States. He acquired an advanced degree and became a successful school administrator, although his marriage had fallen apart. He was devastated when his mother died. Yet he managed to step outside his immediate troubles to assess the situation with a detachment and graceful calm that helped him cope during times of adversity. "I've had as much bad things to happen as good things, but I've never allowed any outside force to take possession of my being," he explained. "That means, whenever I had a problem, I went to something wholesome to solve it." One of the "wholesome" things that helped, he said, was bowling.

The popular image of the Wise Man usually does not include a guy in a bowling shirt, but several qualities have emerged again and again in older people like J. who score high on Ardelt's wisdom scale.

They learn from previous negative experiences. They are able to step outside themselves and assess a troubling situation with calm reflection. They recast a crisis as a problem to be addressed, a puzzle to be solved. They take action in situations they can control and accept the inability to do so when matters are outside their control.

All these sound like noble attributes, but the litany of qualities is so squishy that the definition of wisdom begins to resemble a multi-car pileup of platitudes. One person's positive attitude might be another person's form of self-delusion; perceiving one's limitations might be another name for passivity or indecision or lack of persistence. The common-sense language of wisdom often teeters between proverb and cliché. In fact, the Berlin group mounted an extensive study of proverbs as a way of thinking about wisdom, and Ardelt cites the well-known serenity prayer as an example of a proverb that emphasizes the discernment implicit in wisdom. (This is the saying that goes, "God grant me the serenity to accept the things I cannot change; the courage to change the things I can; and the wisdom to know the difference.")

But as I read the undeniably self-satisfied profiles in wisdom published by Ardelt, they reminded me that wisdom unfolds on many stages and very much depends on the dramatis personae. We tend to think of wisdom as a Cecil B. DeMille production in human enlightenment, with Biblical sets and King Solomon (or some similarly commanding figure) talking down to us from a position of social and moral authority. But in our daily negotiation with the improvident turns of an imperfect world, we probably need a more personal form of wisdom in dealing with in-laws or coping with financial stresses. Perhaps the most important yin-yang of wisdom may be the different shapes it takes in the public and private domains. The public face of wisdom has to do with leadership, judgment, and a responsibility to the collective future, offering a kind of moral inspiration to do the greatest good for the greatest number of people; this is the face that emerges when people are asked in surveys to name people they consider to be wise (the nominees invariably include people like Martin Luther King Jr., Nelson Mandela, Mother Teresa, and again Gandhi). The private face of wisdom may be Vivian Clayton's father, my parents, your Uncle Myron. By comparison, the example of their wisdom is invisible to all but the inner circle of kin and acquaintances that benefit each day, in myriad specific ways, from the exercise of wisdom.

If nothing else, the 3D-WS studies suggest that a kind of wisdom can arise in ordinary people from unexpected backgrounds. With Ardelt's help, I had an opportunity to speak with some of the people who ranked high on her wisdom scale. C., it turns out, grew up on a tobacco farm in Kentucky, never finished high school and harbored no greater ambition than to have children. "We're not mountaineers," she told me, "but we are hillbillies."

Ardelt is now testing her wisdom scale on a different population. In collaboration with George E. Vaillant, a Harvard Medical School psychiatrist affiliated with Brigham and Women's Hospital in Boston, she is assessing a group of Harvard University graduates who have been faithfully filling out psychological surveys every two years since they began college in the late 1930s. "I have identified people I consider wise and people I consider relatively low in wisdom," says Ardelt, who is still analyzing the data. People who rated high in wisdom, she adds, were "very generous," both financially and emotionally; among those who rated low in wisdom, "there was this occupation with the self."

ARDELT ACKNOWLEDGES THAT no one really knows what wisdom is. "I like my definition," she says. "The Baltes people like their definition, and Sternberg likes his. There's no agreement on what wisdom is, and that's the fuzzy part. We're not there yet."

The "fuzziness" of wisdom studies scares many people away from the subject; as James Birren and Cheryl Svensson noted recently, the thirteen chapters of Sternberg's 1990 collection *Wisdom* offer thirteen different approaches, and many self-respecting psychologists and neuroscientists fairly flee from the suggestion that they are investigating the biological basis of wisdom. Yet many of the emotional and cognitive traits that rank high on current research agendas—resilience, positivity, expert knowledge systems, cognitive processing, and especially the regulation of emotion—closely overlap with qualities that have been consistently identified by Clayton, Baltes, Ardelt, and other social scientists as crucial to wisdom.

One of the most interesting areas of neuroscience research involves looking at the way people regulate their emotions and how that regulation can change over the course of a lifetime. Laura Carstensen of Stanford University has produced a substantial body of research over the past two decades showing that the ability to focus on emotional control is tightly linked to a person's sense of time and that older people in general seem to have a better feel for keeping their emotions in balance. This has emerged in part from a long-running research project known informally at Stanford as the "beeper study."

In 1994, Carstensen and her colleagues provided nearly two hundred northern California residents, young and old, with electronic pagers; since then, in several waves of data collection, the subjects have been beeped at random times, up to five times a day over the course of a week, and asked to describe the emotions they are feeling at that exact moment. For Jan Post, who lives north of San Francisco, several of these beeps arrived when she was, as she put it, "doing what husbands and wives are supposed to do." Daniel Zucker's pager pulsed on occasion when he was in meetings at work or driving on the highway. Whatever they were doing, the subjects paused to fill out a questionnaire reporting the intensity of nineteen emotions ranging from anger to happiness to boredom. As part of the ongoing study, participants are now coming into the Stanford lab for intense psychological testing, which often includes a session in brain-scanning machines.

What the Stanford researchers have found—in the laboratory and out in the world—is that despite the well-documented cognitive declines associated with advancing age, older people seem to have figured out how to manage their emotions in a profoundly important way. Compared with younger people, they experience negative emotions less frequently, exercise better control over their emotions, and rely on a complex and nuanced emotional thermostat that allows them to bounce back quickly from adverse moments. Indeed, they typically strive for emotional balance, which in turn seems to affect the ways their brains process information from their environment.

On a recent spring day in Palo Alto, California, for example, the Stanford researchers put sixty-seven-year-old N., a very nice, good-natured subject of the beeper study, through a battery of cognitive and emotional assessments. She repeatedly filled out questionnaires asking her to gauge the intensity of her emotions; took a vocabulary test; endured a wearying series of tasks designed to assess the quality of her memory; and before the two-day gantlet of testing was done, would also undergo functional magnetic resonance imaging (fMRI) of her brain while she performed a monetary-reward task and viewed pictures laden with positive and negative emotional content. Every once in a while, she was asked to chew on a piece of cotton until it was saturated with saliva (a test for the stress hormone cortisol).

These laboratory sessions are not without their frustrating moments, and the low point for N. occurred in the middle of a Tuesday afternoon, when she was asked to perform two different tasks: public speaking and a maddening mathematical task that involved a formula for counting backward as fast as she could. Every time N. made a mistake, and she made quite a few, a humorless examiner would say, "Error," and ask her to start again. She became so flustered that she'd pretzeled her body into an ampersand and kept repeating, "Gosh, I can't even think. . . ." Later she confided, "I was almost in tears right after doing those numbers." But by the time N. completed the final task of the day, which asked her to rate her emotions on a scale of one (for low) to seven (for high), she appeared to have rebounded quite nicely.

"Happiness is a seven," she said with a triumphant laugh, checking the last box on the questionnaire. "I'm getting out of here!"

That, in a sense, is the take-home message of the "beeper study," too. The results suggest that older people on average are more even-keeled and resilient emotionally. "Younger people tend to be either positive or negative at any given point in their daily life," Carstensen says, "but older people are more likely to experience mixed emotions, happiness, and a touch of sadness at the same time. Having mixed emotions helps to regulate emotional states better than extremes of emotion. There's a lot of loss associated with aging, and humans are the only species that recognizes that time eventually runs out. That influences the motivation to savor the day-to-day experiences you have, it allows you to be more positive. Appreciating the fragility of life helps you savor it." Fredda Blanchard-Fields of the Georgia Institute of Technology has produced a series of studies showing that the emotional equilibrium of older people allows them to negotiate solutions to interpersonal problems better than younger people. "She wouldn't call it research on wisdom," Carstensen says of Blanchard-Fields, "but I would."

Carstensen and her colleagues believe that this motivation to focus less on the negative is probably unconscious and shaped by one's sense of time. "According to our theory, this isn't a quality of aging per se, but of time horizons," she says. "When your time perspective shortens, as it does when you come closer to the ends of things, you tend to focus on emotionally meaningful goals. When the time horizon is long, you focus on knowledge acquisition." As time horizons shorten, she added, "things become much clearer, because people are letting their feelings navigate what they do, who they spend time with, what are the choices they're making in life, and it's about right now."

Carstensen calls this "socioemotional selectivity theory" and says that in the shortened time perspective of old age, people are motivated to focus on the positive in a way that registers as a difference in cognitive processing in the brain. "I'm not a 'wisdom person,'" she said in a recent conversation in her office. But she readily agreed that many elements of emotional regulation seen in older adults are "absolutely" consistent with qualities that have long been identified by the wisdom researchers.

This is all of a piece with life-span development theory (Carstensen got her Ph.D. in a program founded by Paul Baltes), which has as a central precept the idea that the decisions one makes at each stage of life involve trade-offs. As Carstensen puts it, "There's always a cost, always a tension, between selecting any goal." She and her colleague Corinna E. Lackenhoff have speculated that there may even be good evolutionary reasons for this division between knowledge acquisition and emotional fulfillment. Acquiring knowledge (and paying close attention to threat and danger) increases the likelihood that young people will survive to reproductive age; emphasizing emotional connection and kinship at an older age may increase the survival ability of one's children and grandchildren (and their genes) in the future. "If you invest increasingly in those people related to you," Carstensen says, "then you are investing in your own genes' survival."

This "positivity" effect may even have long-term health consequences. Although the findings haven't been peer-reviewed or published, Carstensen said preliminary results from the small sample in the ongoing "beeper" experiment indicate that people who didn't regulate their emotions well as adults and were relatively more negative at the start of the study "were more likely to be dead" ten years later, independent of their health status at the beginning of the experiment.

This intriguing correlation between positivity and longevity has been seen elsewhere. In 2002, Becca Levy, a psychologist at Yale University, collaborated with researchers for the Ohio Longitudinal Study, who have been following aging in a cohort of people since 1975, and they made a very surprising finding: older people with a more positive attitude toward old age lived seven and a half years longer. "It's a pretty robust effect," says Suzanne Kunkel, the gerontologist who heads the Ohio study. "People with a positive perception of aging, of themselves as an aging person, seem to have a longevity advantage." But there may also be downsides to positivity, and Carstensen's lab is investigating that possibility. Older people who are inclined to tune out the negative and focus on the positive, she says, might be more vulnerable to confidence scams and make bad, overly trusting decisions.

Richard J. Davidson, a neuroscientist at the University of Wisconsin, has been looking at patterns of brain activity associated with emotional regulation in a small group of older people who have participated in the Wisconsin Longitudinal Study. In a paper published last year, the Wisconsin team reported that older adults (the average age was sixty-four) who regulated their emotions well showed a distinctly different pattern of brain activity than those who didn't. These people apparently used their prefrontal cortex, the part of the brain that exerts "executive control" over certain brain functions, to tamp down activity in the amygdala, a small region deep in the brain that processes emotional content, especially fear and anxiety. In people who are poor regulators of emotion, activity in the amygdala is higher, and daily measurements of the stress hormone cortisol follow a pattern that has been associated with poor health outcomes.

"Those people who are good at regulating negative emotion, inferred by their ability to voluntarily use cognitive strategies to reappraise a stimulus, show reductions in activation in the amygdala," says Davidson, who added that such regulation probably results from "something that has been at least implicitly trained over the years." It is difficult (not to say dangerous) to generalize from such a small, focused study, but the implication is that people who learn, or somehow train themselves, to modulate their emotions are better able to manage stress and bounce back from adversity. Although they can register the negative, they have somehow learned not to get bogged down in it. Whether this learning is a form of "wisdom" accumulated over a lifetime of experience, as wisdom researchers see it, or can be acquired through training exercises like meditation, as Davidson's previous research has shown, the recent message from neuroscience laboratories is that the optimal regulation of emotion can be seen in the brain.

Similarly, several years ago, Carstensen; Mara Mather of the University of California, Santa Cruz; John Gabrieli, a neuroscientist now at the Massachusetts Institute of Technology; and several colleagues performed fMRI studies of young and old people to see whether the ability to regulate emotions left a trace in the amygdala. The study indicated that the amygdala in young people becomes active when they view both positive and negative images; the amygdala in older people is active only when they view positive images. Put another way, young people tend to cling to the negative information, neurologically speaking, while older people seem better able to shrug it off and focus more on positive images. This neural selectivity, this focus on the positive, is virtually instantaneous, Gabrieli says, and yet probably reflects a kind of emotional knowledge or experience that guides cognitive focus; Carstensen says older people "disattend" negative information. This "disattention" also echoes some very old thoughts on wisdom. In his 1890 book *The Principles of Psychology*, William James observed, "The art of being wise is the art of knowing what to overlook." In modern neuroscience parlance, Gabrieli says, "you could say that in older people the amygdala is overlooking the negative."

MUCH OF THE RESEARCH to date has reflected a predominantly Western notion of wisdom, but its definition can be further muddied by cultural vagaries. In one cross-cultural study, researchers found that Americans and Australians essentially equated being wise with being experienced and knowledgeable; being old and discreet were seen as less-than-desirable qualities. People in India and Japan, by contrast, linked wisdom to being discreet, aged, and experienced.

Nevertheless, the notion of wisdom is sufficiently universal that it raises other questions: Where does it come from, and how does one acquire it? Surprisingly, a good deal of evidence, both anecdotal and empirical, suggests that the seeds of wisdom are planted earlier in life—certainly earlier than old age, often earlier than middle age, and possibly even earlier than young adulthood. And there are strong hints that wisdom is associated with an earlier exposure to adversity or failure. That certainly seems to be the case with emotional regulation and is perfectly consistent with Carstensen's ideas about shifting time horizons. Karen Parker and her colleagues at Stanford have published several striking animal studies showing that a very early exposure to mild adversity (she calls it a "stress inoculation") seems to "enhance the development of brain systems that regulate emotional, neuroendocrine and cognitive control"—at least in nonhuman primates. Some researchers are also exploring the genetic basis of resilience.

The Berlin group reported that the roots of wisdom can be traced, in some cases, to adolescence. Jacqui Smith points out that many of the people in the Berlin Aging Study survived two world wars and a global depression; the elderly people who scored high on Monika Ardelt's wisdom scale also reported considerable hardship earlier in their lives.

This notion that wise people might have been "vaccinated" earlier in life by adversity reminded me of Vivian Clayton's father, sitting next to his frail mother in London while the German bombs rained down around them, celebrating their survival each time with a cup of tea. It also made me curious about Clayton, who disappeared from academia in 1981. I managed to track her down through a short item on the Internet, which described a psychologist of the same name who tended bees as a hobby in northern California. It turned out to be the same Vivian Clayton, and she agreed to meet with me at her office in Orinda on a sunny March morning, a few hours before seeing her first patient of the day.

Now fifty-six—"and proud of it," she said—Clayton turned out to be a vivacious woman with a soothingly enthusiastic voice. After all the abstraction involved in thinking about wisdom, she had turned to a more pragmatic role as a geriatric neuropsychologist, helping families and lawyers determine mental capacity in older people experiencing cognitive declines; in fact, she helped write the California State Bar manual for making these determinations. She never contributed anything to the wisdom field after 1982, although Paul Baltes continued to send her papers from Berlin and Monika Ardelt has occasionally sought her counsel. I asked her if she regretted not continuing in the field, and she said not at all. "I reached a fork in the road," she said. "Wisdom can be a very abstract concept, and as I got older, I gravitated to more practical approaches."

We talked about wisdom in contemporary culture, and gradually the conversation turned to bees. "You know, bees have been around for hundreds of millions of years, at least, as living creatures," Clayton said. "And when you work a hive, and you're there with that hive alone, and you hear how contented the bees are, you just have the sense that they have the pulse of the universe encoded in their genes. And I really feel that the concept of wisdom is like that, too. Somehow, like the bees, we are programmed to understand when someone has been wise. But what wisdom is, and how one learns to be wise, is still somewhat of a mystery."

AL GORE

Moving Beyond Kyoto

FROM THE *NEW YORK TIMES*

In a year that saw former U.S. vice president Al Gore win an Oscar and a Nobel Prize, he still found other ways to raise awareness of what he deems our "planetary emergency," the threat of global warming—such as the Live Earth concert, and this essay.

W E—THE HUMAN SPECIES—HAVE arrived at a moment of decision. It is unprecedented and even laughable for us to imagine that we could actually make a conscious choice as a species, but that is nevertheless the challenge that is before us.

Our home—Earth—is in danger. What is at risk of being destroyed is not the planet itself, but the conditions that have made it hospitable for human beings.

Without realizing the consequences of our actions, we have begun to put so much carbon dioxide into the thin shell of air surrounding our world that we have literally changed the heat balance between Earth and the Sun. If we don't stop doing this pretty quickly, the average temperature will increase to levels humans have never known and put an end to the favorable climate balance on which our civilization depends.

In the last 150 years, in an accelerating frenzy, we have been removing increasing quantities of carbon from the ground—mainly in the form of coal and oil—and burning it in ways that dump seventy million tons of CO_2 every twenty-four hours into the Earth's atmosphere.

The concentrations of CO_2—having never risen above 300 parts per million for at least a million years—have been driven from 280 parts per million at the beginning of the coal boom to 383 parts per million this year.

As a direct result, many scientists are now warning that we are moving closer to several "tipping points" that could—within ten years—make it impossible for us to avoid irretrievable damage to the planet's habitability for human civilization.

Just in the last few months, new studies have shown that the north polar ice cap—which helps the planet cool itself—is melting nearly three times faster than the most pessimistic computer models predicted. Unless we take action, summer ice could be completely gone in as little as thirty-five years. Similarly, at the other end of the planet, near the South Pole, scientists have found new evidence of snow melting in West Antarctica across an area as large as California.

This is not a political issue. This is a moral issue, one that affects the survival of human civilization. It is not a question of left versus right; it is a question of right versus wrong. Put simply, it is wrong to destroy the habitability of our planet and ruin the prospects of every generation that follows ours.

On September 21, 1987, President Ronald Reagan said, "In our obsession with antagonisms of the moment, we often forget how much unites all the members of humanity. Perhaps we need some outside, universal threat to recognize this common bond. I occasionally think how quickly our differences would vanish if we were facing an alien threat from outside this world."

We—all of us—now face a universal threat. Though it is not from outside this world, it is nevertheless cosmic in scale.

Consider this tale of two planets. Earth and Venus are almost exactly the same size, and have almost exactly the same amount of carbon. The difference is that most of the carbon on Earth is in the ground—having been deposited there by various forms of life over the last six hundred million years—and most of the carbon on Venus is in the atmosphere.

As a result, while the average temperature on Earth is a pleasant 59 degrees, the average temperature on Venus is 867 degrees. True, Venus is closer to the Sun than we are, but the fault is not in our star; Venus is three times hotter on average than Mercury, which is right next to the Sun. It's the carbon dioxide.

This threat also requires us, in Reagan's phrase, to unite in recognition of our common bond.

Next Saturday, on all seven continents, the Live Earth concert will ask for the attention of humankind to begin a three-year campaign to make everyone on our planet aware of how we can solve the climate crisis in time to avoid catastrophe. Individuals must be a part of the solution. In the words of Buckminster Fuller, "If the success or failure of this planet, and of human beings, depended on how I am and what I do, how would I be? What would I do?"

Live Earth will offer an answer to this question by asking everyone who attends or listens to the concerts to sign a personal pledge to take specific steps to combat climate change. (More details about the pledge are available at algore.com.)

But individual action will also have to shape and drive government action. Here Americans have a special responsibility. Throughout most of our short history, the United States and the American people have provided moral leadership for the world. Establishing the Bill of Rights, framing democracy in the Constitution, defeating fascism in World War II, toppling Communism, and landing on the moon—all were the result of American leadership.

Once again, Americans must come together and direct our government to take on a global challenge. American leadership is a precondition for success.

To this end, we should demand that the United States join an international treaty within the next two years that cuts global warming pollution by 90 percent in developed countries and by more than half worldwide in time for the next generation to inherit a healthy Earth.

This treaty would mark a new effort. I am proud of my role during the Clinton administration in negotiating the Kyoto protocol. But I believe that the protocol has been so demonized in the United States that it probably cannot be ratified here—much in the way the Carter administration was prevented from winning ratification of an expanded strategic arms limitation treaty in 1979. Moreover, the negotiations will soon begin on a tougher climate treaty.

Therefore, just as President Reagan renamed and modified the SALT agreement (calling it Start), after belatedly recognizing the need for it, our next president must immediately focus on quickly concluding a new and even tougher climate change pact. We should aim to complete this global treaty by the end of 2009—and not wait until 2012 as currently planned.

If by the beginning of 2009, the United States already has in place a domestic regime to reduce global warming pollution, I have no doubt that when we give industry a goal and the tools and flexibility to sharply reduce carbon emissions, we can complete and ratify a new treaty quickly. It is, after all, a planetary emergency.

A new treaty will still have differentiated commitments, of course; countries will be asked to meet different requirements based upon their historical share or contribution to the problem and their relative ability to carry the burden of change. This precedent is well established in international law, and there is no other way to do it.

There are some who will try to pervert this precedent and use xenophobia or nativist arguments to say that every country should be held to the same standard. But should countries with one-fifth our gross domestic product—countries that contributed almost nothing in the past to the creation of this crisis—really carry the same load as the United States? Are we so scared of this challenge that we cannot lead?

Our children have a right to hold us to a higher standard when their future—indeed, the future of all human civilization—is hanging in the balance. They deserve better than a government that censors the best scientific evidence and harasses honest scientists who try to warn us about looming catastrophe. They deserve better than politicians who sit on their hands and do nothing to confront the greatest challenge that humankind has ever faced—even as the danger bears down on us.

We should focus instead on the opportunities that are part of this challenge. Certainly, there will be new jobs and new profits as corporations move aggressively to capture the enormous economic opportunities offered by a clean energy future.

But there's something even more precious to be gained if we do the right thing. The climate crisis offers us the chance to experience what few generations in history have had the privilege of experiencing: a generational mission; a compelling moral purpose; a shared cause; and the thrill of being forced by circumstances to put aside the pettiness and conflict of politics and to embrace a genuine moral and spiritual challenge.

JIM YARDLEY

Beneath Booming Cities, China's Future Is Drying Up

FROM THE *NEW YORK TIMES*

Among the stresses on the world's environment, perhaps one of the most alarming is the ecological crisis that is being created by China's economic growth. Jim Yardley reports that partly as a result of that runaway growth, China is running out of water.

SHIJIAZHUANG, CHINA—HUNDREDS OF FEET below ground, the primary water source for this provincial capital of more than two million people is steadily running dry. The underground water table is sinking about four feet a year. Municipal wells have already drained two-thirds of the local groundwater.

Above ground, this city in the North China Plain is having a party. Economic growth topped 11 percent last year. Population is rising. A new upscale housing development is advertising waterfront property on lakes filled with pumped groundwater. Another half-built complex, the Arc de Royal, is rising above one of the lowest points in the city's water table.

"People who are buying apartments aren't thinking about whether there will be water in the future," said Zhang Zhongmin, who has tried for twenty years to raise public awareness about the city's dire water situation.

For three decades, water has been indispensable in sustaining the rollicking economic expansion that has made China a world power. Now, China's galloping, often wasteful style of economic growth is pushing the country toward a water crisis. Water pollution is rampant nationwide, while water scarcity has worsened severely in north China—even as demand keeps rising everywhere.

China is scouring the world for oil, natural gas, and minerals to keep its economic machine humming. But trade deals cannot solve water problems. Water usage in China has quintupled since 1949, and leaders will increasingly face tough political choices as cities, industry, and farming compete for a finite and unbalanced water supply.

One example is grain. The Communist Party, leery of depending on imports to feed the country, has long insisted on grain self-sufficiency. But growing so much grain consumes huge amounts of underground water in the North China Plain, which produces half the country's wheat. Some scientists say farming in the rapidly urbanizing region should be restricted to protect endangered aquifers. Yet doing so could threaten the livelihoods of millions of farmers and cause a spike in international grain prices.

For the Communist Party, the immediate challenge is the prosaic task of forcing the world's most dynamic economy to conserve and protect clean water. Water pollution is so widespread that regulators say a major incident occurs every other day. Municipal and industrial dumping has left sections of many rivers "unfit for human contact."

Cities like Beijing and Tianjin have shown progress on water conservation, but China's economy continues to emphasize growth. Industry in China uses three to ten times more water, depending on the product, than industries in developed nations.

"We have to now focus on conservation," said Ma Jun, a prominent environmentalist. "We don't have much extra water resources. We have the same resources and much bigger pressures from growth."

In the past, the Communist Party has reflexively turned to engineering projects to address water problems, and now it is reaching back to one of Mao's unrealized plans: the sixty-two-billion-dollar South-to-North Water Transfer Project to funnel more than twelve trillion gallons northward every year along three routes from the Yangtze River basin, where water is more abundant. The project, if fully built, would be completed in 2050. The eastern and central lines are already under construction; the western line, the most disputed because of environmental concerns, remains in the planning stages.

The North China Plain undoubtedly needs any water it can get. An economic powerhouse with more than two hundred million people, it has limited rainfall and depends on groundwater for 60 percent of its supply. Other countries, like Yemen, India, Mexico, and the United States, have aquifers that are being drained to dangerously low levels. But scientists say those below the North China Plain may be drained within thirty years.

"There's no uncertainty," said Richard Evans, a hydrologist who has worked in China for two decades and has served as a consultant to the World Bank and China's Ministry of Water Resources. "The rate of decline is very clear, very well documented. They will run out of groundwater if the current rate continues."

Outside Shijiazhuang, construction crews are working on the transfer project's central line, which will provide the city with infusions of water on the way to the final destination, Beijing. For many of the engineers and workers, the job carries a patriotic gloss.

Yet while many scientists agree that the project will provide an important influx of water, they also say it will not be a cure-all. No one knows how much clean water the project will deliver; pollution problems are already arising on the eastern line. Cities and industry will be the beneficiaries of the new water, but the impact on farming is limited. Water deficits are expected to remain.

"Many people are asking the question: What can they do?" said Zheng Chunmiao, a leading international groundwater expert. "They just cannot continue with current practices. They have to find a way to bring the problem under control."

A DRYING REGION

On a drizzly, polluted morning last April, Wang Baosheng steered his Chinese-made sport utility vehicle out of a shopping center on the west side of Beijing for a three-hour southbound commute that became a tour of the water crisis on the North China Plain.

Mr. Wang travels several times a month to Shijiazhuang, where he is chief engineer overseeing construction of three miles of the central line of the water transfer project. A light rain splattered the windshield, and he recited a Chinese proverb about the preciousness of spring showers for farmers. He also noticed one dead river after another as his SUV glided over dusty, barren riverbeds: the Yongding, the Yishui, the Xia, and, finally, the Hutuo. "You see all these streams with bridges, but there is no water," he said.

A century or so ago, the North China Plain was a healthy ecosystem, scientists say. Farmers digging wells could strike water within eight feet. Streams and creeks meandered through the region. Swamps, natural springs, and wetlands were common.

Today, the region, comparable in size to New Mexico, is parched. Roughly five-sixths of the wetlands have dried up, according to one study. Scientists say that most natural streams or creeks have disappeared. Several rivers that once were navigable are now mostly dust and brush. The largest natural freshwater lake in northern China, Lake Baiyangdian, is steadily contracting and besieged with pollution.

What happened? The list includes misguided policies, unintended consequences, a population explosion, climate change and, most of all, relentless economic growth. In 1963, a flood paralyzed the region, prompting Mao to construct a flood-control system of dams, reservoirs, and concrete spillways. Flood control improved but the ecological balance was altered as the dams began choking off rivers that once flowed eastward into the North China Plain.

The new reservoirs gradually became major water suppliers for growing cities like Shijiazhuang. Farmers, the region's biggest water users, began depending almost exclusively on wells. Rainfall steadily declined in what some scientists now believe is a consequence of climate change.

Before, farmers had compensated for the region's limited annual rainfall by planting only three crops every two years. But underground water seemed limitless and government policies pushed for higher production, so farmers began planting a second annual crop, usually winter wheat, which requires a lot of water.

By the 1970s, studies show, the water table was already falling. Then Mao's death and the introduction of market-driven economic reforms spurred a farming renaissance. Production soared, and rural incomes rose. The water table kept falling, further drying out wetlands and rivers.

Around 1900, Shijiazhuang was a collection of farming villages. By 1950, the population had reached 335,000. This year, the city has roughly 2.3 million people with a metropolitan area population of 9 million.

More people meant more demand for water, and the city now heavily pumps groundwater. The water table is falling more than a meter a year. Today, some city wells must descend more than six hundred feet to reach clean water. In the deepest drilling areas, steep downward funnels have formed in the water table that are known as "cones of depression."

Groundwater quality also has worsened. Wastewater, often untreated, is now routinely dumped into rivers and open channels. Mr. Zheng, the water specialist, said studies showed that roughly three-quarters of the region's entire aquifer system was now suffering some level of contamination.

"There will be no sustainable development in the future if there is no groundwater supply," said Liu Changming, a leading Chinese hydrology expert and a senior scholar at the Chinese Academy of Sciences.

A NATIONAL PROJECT

Three decades ago, when Deng Xiaoping shifted China from Maoist ideology and fixated the country on economic growth, a generation of technocrats gradually took power and began rebuilding a country that ideology had almost destroyed. Today, the top leaders of the Communist Party—including Hu Jintao, China's president and party chief—were trained as engineers.

Though not members of the political elite, Wang Baosheng, the engineer on the water transfer project, and his colleague Yang Guangjie are of the same background. This spring, at the site outside Shijiazhuang, bulldozers clawed at a V-shaped cut in the dirt while teams of workers in blue jumpsuits and orange hard hats smoothed wet cement over a channel that will be almost as wide as a football field.

"I've been to the Hoover Dam, and I really admire the people who built that," said Mr. Yang, the project manager. "At the time, they were making a huge contribution to the development of their country."

He compared China's transfer project to the water diversion system devised for southern California in the last century. "Maybe we are like America in the 1920s and 1930s," he said. "We're building the country."

China's disadvantage, compared with the United States, is that it has a smaller water supply yet almost five times as many people. China has about 7 percent of the world's water resources and roughly 20 percent of its population. It also has a severe regional water imbalance, with about four-fifths of the water supply in the south.

Mao's vision of borrowing water from the Yangtze for the north had an almost profound simplicity, but engineers and scientists spent decades debating the project before the government approved it, partly out of desperation, in 2002. Today, demand is far greater in the north, and water quality has badly deteriorated in the south. Roughly 41 percent of China's wastewater is now dumped in the Yangtze, raising concerns that siphoning away clean water northward will exacerbate pollution problems in the south.

The upper reaches of the central line are expected to be finished in time to provide water to Beijing for the Olympic Games next year. Mr. Evans, the World Bank consultant, called the complete project "essential" but added that success would depend on avoiding waste and efficiently distributing the water.

Mr. Liu, the scholar and hydrologist, said that farming would get none of the new water and that cities and industry must quickly improve wastewater treatment. Otherwise, he said, cities will use the new water to dump more polluted wastewater. Shijiazhuang now dumps untreated wastewater into a canal that local farmers use to irrigate fields.

For years, Chinese officials thought irrigation efficiency was the answer for reversing groundwater declines. Eloise Kendy, a hydrology expert with The Nature Conservancy who has studied the North China Plain, said that farmers had made improvements but that the water table had kept sinking. Ms. Kendy said the spilled water previously considered "wasted" had actually soaked into the soil and recharged the aquifer. Efficiency erased that recharge. Farmers also used efficiency gains to irrigate more land.

Ms. Kendy said scientists had discovered that the water table was dropping because of water lost by evaporation and transpiration from the soil, plants, and leaves. This lost water is a major reason the water table keeps dropping, scientists say.

Farmers have no choice. They drill deeper.

DIFFICULT CHOICES AHEAD

For many people living in the North China Plain, the notion of a water crisis seems distant. No one is crawling across a parched desert in search of an oasis. But every year, the water table keeps dropping. Nationally, groundwater usage has almost doubled since 1970 and now accounts for one-fifth of the country's total water usage, according to the China Geological Survey Bureau.

The Communist Party is fully aware of the problems. A new water pollution law is under consideration that would sharply increase fines against polluters. Different coastal cities are building desalination plants. Multinational waste treatment companies are being recruited to help tackle the enormous wastewater problem.

Many scientists believe that huge gains can still be reaped by better efficiency and conservation. In north China, pilot projects are under way to try to reduce water loss from winter wheat crops. Some cities have raised the price of water to promote conservation, but it remains subsidized in most places. Already, some cities along the route of the transfer project are recoiling because of the planned higher prices. Some say they may just continue pumping.

Tough political choices, though, seem unavoidable. Studies by different scientists have concluded that the rising water demands in the North China Plain make it unfeasible for farmers to continue planting a winter crop. The international ramifications would be significant if China became an ever bigger customer on world grain markets. Some analysts have long warned that grain prices could steadily rise, contributing to inflation and making it harder for other developing countries to buy food.

The social implications would also be significant inside China. Near Shijiazhuang, Wang Jingyan's farming village depends on wells that are more than six hundred feet deep. Not planting winter wheat would amount to economic suicide.

"We would lose 60 percent or 70 percent of our income if we didn't plant winter wheat," Mr. Wang said. "Everyone here plants winter wheat."

Another water proposal is also radical: huge, rapid urbanization. Scientists say converting farmland into urban areas would save enough water to stop the drop in the water table, if not reverse it, because widespread farming still uses more water than urban areas. Of course, large-scale urbanization, already under way, could worsen air quality; Shijiazhuang's air already ranks among the worst in China because of heavy industrial pollution.

For now, Shijiazhuang's priority, like that of other major Chinese cities, is to grow as quickly as possible. The city's gross domestic product has risen by an average of 10 percent every year since 1980, even as the city's per capita rate of available water is now only one-thirty-third of the world average.

"We have a water shortage, but we have to develop," said Wang Yongli, a senior engineer with the city's water conservation bureau. "And development is going to be put first."

Mr. Wang has spent four decades charting the steady extinction of the North China Plain's aquifer. Water in Shijiazhuang, with more than eight hundred illegal wells, is as scarce as it is in Israel, he said. "In Israel, people regard water as more important than life itself," he said. "In Shijiazhuang, it's not that way. People are focused on the economy."

JOSEPH KAHN

In China, a Lake's Champion Imperils Himself

FROM THE *NEW YORK TIMES*

Because of the closeness between government officials and business leaders in China, rampant pollution and other ravages of the environment can go unchecked. Joseph Kahn profiles one man who tried to sound the alarm about how chemical factories were destroying a beloved lake—and the price he paid for speaking out.

ZHOUTIE, CHINA—LAKE TAI, the center of China's ancient "land of fish and rice," succumbed this year to floods of industrial and agricultural waste.

Toxic cyanobacteria, commonly referred to as pond scum, turned the big lake fluorescent green. The stench of decay choked anyone who came within a mile of its shores. At least two million people who live amid the canals, rice paddies, and chemical plants around the lake had to stop drinking or cooking with their main source of water.

The outbreak confirmed the claims of a crusading peasant, Wu Lihong, who protested for more than a decade that the region's thriving chemical industry, and its powerful friends in the local government, were destroying one of China's ecological treasures.

Mr. Wu, however, bore silent witness. Shortly before the algae crisis erupted in May, the authorities here in his hometown arrested him. In mid-August, with a fetid smell still wafting off the lake, a local court sentenced him to three years on an alchemy of charges that smacked of official retribution.

Pollution has reached epidemic proportions in China, in part because the ruling Communist Party still treats environmental advocates as bigger threats than the degradation of air, water, and soil that prompts them to speak out.

Senior officials have tried to address environmental woes mostly through pulling the traditional levers of China's authoritarian system: issuing command quotas on energy efficiency and emissions reduction; punishing corrupt officials who shield polluters; planting billions of trees across the country to hold back deserts and absorb carbon dioxide.

But they do not dare to unleash individuals who want to make China cleaner. Grass-roots environmentalists arguably do more to expose abuses than any edict emanating from Beijing. But they face a political climate that varies from lukewarm tolerance to icy suppression.

Fixing the environment is, in other words, a political problem. Central party officials say they need people to report polluters and hold local governments to account. They granted legal status to private citizens' groups in 1994 and have allowed environmentalism to emerge as an incipient social force.

But local officials in China get ahead mainly by generating high rates of economic growth and ensuring social order. They have wide latitude to achieve those goals, including nearly complete control over the police and the courts in their domains. They have little enthusiasm for environmentalists who appeal over their heads to higher-ups in the capital.

Mr. Wu, a jaunty, forty-year-old former factory salesman, pioneered a style of intrepid, media-savvy environmental work that made Lake Tai, and the hundreds of chemical factories on its shores, the focus of intense regulatory scrutiny.

In 2005 he was declared an "Environmental Warrior" by the National People's Congress. His address book contained cell phone numbers for officials in Beijing and the provincial capital of Nanjing who outranked the party bosses where he lived.

But Mr. Wu was far from untouchable. He lost his job. His wife lost hers. The police summoned, detained, and interrogated him. The local government and factory owners also tried for years to bring him into the fold with contracts, gifts, and jobs. When party officials offered him a chance to profit handsomely from a pollution cleanup contract, a friend warned him not to accept. Mr. Wu, who needed the money, said yes.

LAKE OF PLENTY

The country's third largest freshwater body, Lake Tai, or Taihu in Chinese, has long provided the people of the lower Yangtze River Delta with both their wealth and their conception of natural beauty.

It nurtured a bounty of the "three whites," white shrimp, whitebait, and whitefish, and a freshwater crustacean delicacy called the hairy crab. Natural and man-made streams irrigated rice paddies, and a network of canals ferried that produce far and wide.

Along the lake's northern reaches, near the city of Wuxi, placid waters and misty hills captured the imagination of Chinese for hundreds of years. The wealthy built gardens that featured the lake's wrinkled, water-scarred limestone rocks set in groves of bamboo and chrysanthemum.

Since the 1950s, however, Lake Tai has been under assault. The authorities constructed dams and weirs to improve irrigation and control floods, disrupting the cleansing circulation of fresh water. Phosphates and other pollution-borne nutrients made the lake eutrophic, sucking out oxygen that fish need to survive.

Even in its degraded state, Lake Tai made an ideal habitat for China's chemical industry, which expanded prolifically in the 1980s. Chemical factories consume and discharge large quantities of water, which the lake provided and absorbed. Its canals made it easy to ship goods to the big industrial port city of Shanghai, downstream.

With strong local government support, the northern arc of Lake Tai became home to twenty-eight hundred chemical plants, most of them small cinder-block factories that took over rice paddies beside canals.

Mr. Wu's hometown alone had three hundred such plants. His narrow village road was reinforced with concrete to withstand the weight of cargo trucks. Factories here made food additives, solvents, and adhesives.

The industry transformed the economy. By the mid-1990s, taxes on chemical industry profits accounted for four-fifths of local government revenue, according to a report from the city of Yixing, which oversees Zhoutie.

Mr. Wu benefited as well. In his early twenties, he got a salaried job as salesman for a factory that made soundproofing material. It allowed him to travel around the country, and paid nice commissions on his sales. His wife, Xu Jiehua, made dyes.

Mr. Wu took long walks after dinner. The acrid tinge in the cool night air was the smell of prosperity to some locals. But it nauseated him, Mr. Wu recalled in later interviews.

In streams where he and Ms. Xu played as children, teeming whitefish used to tickle their legs. By the early 1990s, there were no fish in the streams, which ran black and red. "Rivers of blood," Ms. Xu quoted him as saying.

Mr. Wu is small and pudgy. Ms. Xu calls him "little fatty." He also has a short temper, and pollution sparked it.

"In the beginning I didn't understand it myself," he recalled years later in an interview with *Farmers' Daily*. "It was my personality that decided all of this. I felt the burden getting bigger."

He began by snapping photos of factories dumping untreated effluent into canals. He mailed them, anonymously at first, to environmental protection agencies.

When that produced few results, he signed the letters and included his phone number, volunteering to help inspectors see the problem for themselves.

Local regulators ignored him. But fish kills, declining rice yields, and slumping tourism to the once pristine area made Lake Tai's ecology a broader concern. Higher-ranking officials in Nanjing, capital of Jiangsu Province, got in touch.

One evening, Mr. Wu brought provincial inspectors to see concealed pipes running from a factory near his home to a stream that flowed into the lake. The factory, Feida Chemical, got slapped with a fine, and Mr. Wu got his start.

FRIENDS AND ENEMIES

Mr. Wu's farmhouse filled up with the evidence he amassed, a bit haphazardly, of a looming environmental disaster. He used his pantry to store plastic bottles containing muddy water samples from streams and canals. Near his queen-size bed he kept stacks of newspaper clippings and photographs, letters and petitions.

One letter from local farmers described how a nearby factory making 8-hydroxyquinoline, used as a deodorant and antiseptic, emitted noxious fumes that "make our days and nights impassable." Another writer referred to a local factory as "a new Unit 731," after the Japanese team that conducted chemical warfare experiments in

World War II. Members of another group said they did not dare tend their rice paddies without wearing gloves and galoshes because irrigation water caused their skin to peel off.

Mr. Wu answered many such calls for help. Between 1998 and 2006, the environmental protection agency of Jiangsu Province recorded receiving two hundred reports of pollution incidents and regulatory violations from Mr. Wu.

Many of those he helped became allies. But Mr. Wu was making as many enemies as friends.

"Our society lacks the right atmosphere for environmental protection," he told one local newspaper. "Even in areas where pollution is most severe, I still have a hard time winning people's support."

Some residents feared for their jobs, with good reason. The soundproofing factory fired Mr. Wu in 1999. His notice of dismissal, which he saved among his other papers, cited his failure to attend a meeting.

His family lived off his wife's salary at the dye factory for a time. Then one day Ms. Xu mentioned to Mr. Wu how the stream near her factory changed colors depending on which dye they made that day. Mr. Wu brought a television crew to film the rainbow-colored stream. Ms. Xu soon lost her job as well.

"He did not always have our family's happiness at heart," Ms. Xu recalled. "He probably should have investigated someone else's factory."

Such pressure, though, made him confront local authorities more directly.

In 2001, Wen Jiabao, then a vice premier, now China's prime minister, came to investigate reports of Lake Tai's deterioration. Like most Communist Party inspection tours, word of this one reached local officials in advance. When Mr. Wen asked to see a typical dye plant, one was made ready, according to several people who witnessed the preparations.

The factory got a fresh coat of paint. The canal that ran beside it was drained, dredged and refilled with fresh water. Shortly before

Mr. Wen's motorcade arrived, workers dumped thousands of carp into the canal. Farmers were positioned along the banks holding fishing rods.

Mr. Wen spent twenty minutes there. A picture of him shaking hands with the factory boss hangs in its lobby.

Mr. Wu fired off an angry letter to Beijing recounting the ruse and warning the vice premier that he had been "deceived." Mr. Wu circulated copies among his friends. Local officials saw it, too. Several villagers said they were warned then that they should keep a distance from Mr. Wu.

WORDS FROM ABOVE

One summer afternoon in 2002, Mr. Wu went out on an errand and saw a banner stretched across the main road downtown. It read: "Warmly welcome the police to arrest Wu Lihong for committing blackmail in the name of environmentalism."

Mr. Wu told friends he initially suspected that the banner was hung by local factory bosses to intimidate him. But when he went to the police to complain, he found a stack of placards with the same exhortation in the police station. The police had erected the banner themselves, and they detained him on the spot.

His family received a detention notice accusing Mr. Wu of inciting farmers to stage a public protest about pollution a few weeks earlier. The notice did not mention blackmail, as the banner had, and the police never pressed charges. He was released within two weeks.

That episode appeared to be part of an inconsistent, somewhat bumbling effort to keep Mr. Wu boxed up and harmless.

There were carrots as well as sticks. Zhang Aiguo, the chief environmental regulator in the city of Yixing, struck up a dialogue with Mr. Wu, several friends said.

Hang Yaobin, a truck driver and sundry shop owner in Zhoutie who has also pressed for better environmental controls, said Mr.

Zhang told Mr. Wu that they could improve the environment together. But Mr. Wu should expose problems in other jurisdictions and should stop damaging Yixing's reputation.

"Zhang Aiguo told him: 'Don't make me stink, or I'll lose my job. Then we'll accomplish nothing,'" Mr. Hang said.

In a telephone interview, Mr. Zhang declined to discuss his dealings with Mr. Wu in detail. But he acknowledged that the two talked regularly before he was assigned to another position in the Yixing government.

In 2003, Mr. Zhang offered Mr. Wu a business opportunity. A steel plant in Zhoutie had been ordered by environmental authorities to buy new dust-control equipment. Mr. Wu could find a vendor for the equipment and earn a handsome commission, several people told about the arrangement said.

Mr. Zhang confirmed that he told Mr. Wu of the opportunity.

Mr. Wu debated whether to accept. Mr. Hang said he advised his friend against it. "If you're engaged in a confrontation with officials you can't gamble, or visit prostitutes, or have any other vice," Mr. Han said. "They are always looking for ways to get you."

But this contract involved an environmental cleanup. And with both Mr. Wu and his wife out of work, they needed money. Mr. Wu agreed to contact a vendor recommended by Mr. Zhang.

It was not a rewarding endeavor. He brokered a contract. But the dust-control company gave him only a token advance on his promised commission. The steel plant boss, who had befriended Mr. Wu, eventually withheld part of what he owed the dust-control company to compensate Mr. Wu, according to Ms. Xu, his wife.

That was one of several muddled interactions with local officials and businessmen that did not satisfy either side. Mr. Wu remained cash-strapped. He did not stop contacting Nanjing and Beijing about pollution problems.

In 2005, he heard that the local government would be the host of a big delegation of Chinese reporters as part of the China Environmental Century Tour. He got in touch with China Central Television, the leading national broadcaster, and promised to reveal the story behind the story.

He arranged covertly for the reporters to inspect a section of the Caoqiao River that he learned the government planned to show them on the coming tour. He revealed hidden pipes that discharged black effluent from local factories into the river, which flows into Lake Tai.

The China Central Television crew later joined the Potemkin official tour. They aired a special report on "the river that goes from black to clear overnight."

Mr. Wu was the star of that report, an environmental celebrity. Later the same year, the National People's Congress, China's party-run Parliament, declared him an "Environmental Warrior."

MODEL CITY

With President Hu Jintao and Mr. Wen demanding tougher action on pollution, local officials in 2006 came under new pressure to clean up Lake Tai. Despite repeated pledges and campaigns to protect the once scenic lake, it was still rated Grade V by the State Environmental Protection Administration, the lowest level on its scale.

Yixing ordered a new crackdown on small chemical factories. It claimed to have reduced the total number by half from the peak of twenty-eight hundred in the late 1990s. The city said the industry, which once accounted for as much as 85 percent of the area's industrial output, constituted just 40 percent in 2006.

But local officials put at least as much emphasis on fighting the perception that they had a pollution problem. They lobbied heavily for the State Environmental Protection Administration to declare it a "Model City for Environmental Protection."

Around the same time, Wu Xijun, the Communist Party boss of Zhoutie, called Mr. Wu to his office. The two Mr. Wus, who are not related, had a "face-to-face talk" about the damage Wu Lihong's environmental protests were doing to the area's reputation. The party secretary then made him an offer, according to friends of Mr. Wu and an official court document that confirmed the meeting.

In March 2006, the township party committee paid Mr. Wu to promote tourism on the condition that he stop "nonfactual reporting" of pollution problems. The payments totaled about five thousand dollars, the court document confirmed.

Mr. Wu may have toned down his protests for a time, friends said. But early this year, he learned that Yixing had won the environmental administration's designation as a "Model City for Environmental Protection." Enraged, he began his most assertive effort to date to embarrass local officials.

He spent weeks traveling throughout the area on his motorcycle, collecting water samples and photographing rivers and canals. He gathered data he hoped could prove that factories released most of their polluted water at night in quantities that the currents could wash away by dawn.

In April, he prepared to bring the water samples and photographic evidence to Beijing. He told friends he intended to file a lawsuit there against SEPA, the environmental administration, for its decision to honor Yixing. He never made the trip.

On the night of April 13, several dozen police and state security officers raided his farmhouse. Climbing ladders, they pried open the windows to his second-floor bedroom, arresting him and seizing documents and a computer.

Prosecutors quickly indicted Mr. Wu on two charges of blackmail. The first charge claimed that after he "gained knowledge" of a contract between the steel company and the dust-control company in 2003, he threatened to use his connections to undermine it unless the dust-control company paid him to keep quiet.

The second charge claimed that Mr. Wu extorted money from the Communist Party Committee of Zhoutie by threatening to report pollution problems.

Prosecutors revised the indictment twice in the following weeks. They dropped the charge of blackmailing the Communist

Party, offering no explanation. Then they added a new charge, this one for "fraud." It claimed that Mr. Wu had illegally aided the steel company boss in preparing false documentation to account for the money the steel company paid Mr. Wu in 2003.

The three indictments each claimed that Mr. Wu confessed to the various charges. The last week of May, with Mr. Wu in custody, Lake Tai cried for help. Nitrogen and phosphorous, the untreated residue of chemical processing, fertilizer, and sewage, built up to record levels, while rainfall fell short.

LAKE TAI'S REVENGE

Lake Tai had algal blooms before. This time, according to an analysis by the State Environmental Protection Administration, cyanobacteria "exploded" at rates that had not been seen in the past. Much of the lake was covered with a deep, foul-smelling canopy that left most of the 2.3 million people in Wuxi, the biggest city on the northern part of the lake, without drinking water for many days.

Local officials initially called the outbreak a "natural disaster." But state media rushed to the scene, and some showed pictures of chemical factories dumping waste into the lake even as residents formed long lines at supermarkets to buy bottled water.

Neighboring cities shut sluice gates and canal locks to prevent contamination, creating a monumental maritime traffic jam and further reducing circulation around Lake Tai. The problem did not ease until central authorities ordered Yangtze River water diverted into the lake. Even then, the bloom lingered into late summer.

Mr. Wen convened a meeting of the State Council to discuss the matter. "The pollution of Lake Tai has sounded the alarm for us," state media quoted him as saying. "The problem has never been tackled at its root."

Five party and government officials in Yixing and Zhoutie, including three involved in environmental work, were dismissed or demoted. Li Yuanchao, the party boss of Jiangsu Province, vowed to clean up Lake Tai even if it meant taking a 15 percent cut to the province's economic output. Authorities pledged to shut down hundreds of the most egregious polluters in their most sweeping crackdown to date.

Ms. Xu, Mr. Wu's wife, said she hoped the authorities would conclude that it would be improper, or at least inconvenient, to prosecute Mr. Wu under such conditions. His trial, initially scheduled for June, was delayed, prompting speculation that someone at a higher level had intervened.

But although Mr. Wu's arrest generated attention in both the domestic and international media, there is no indication that central government officials objected to his prosecution. On a Friday afternoon in August, the road in front of Yixing's courthouse filled with Volkswagen Santanas, the standard-issue sedans of China's police and security services. In a park nearby, officials hung a banner advertising the city's new status as a "Model City for Environmental Protection."

The evidence against Mr. Wu consisted mainly of written testimony and his own confession. The judges rejected a request by Mr. Wu's lawyer to summon prosecution witnesses for cross-examination.

Mr. Wu told the judges in open court that the police had deprived him of food and forced him to stay awake for five days and five nights in succession, relenting only when he signed a written confession. He said that the confession was coerced and that he was innocent. The judges ruled that since Mr. Wu could not prove that he had been tortured, his confession remained valid.

Mr. Wu lost his temper. "Since I was a child I have never broken the law," he shouted, according to relatives who attended. "If I could right now, I would like to split you in two." He was sentenced to three years.

Shortly after the trial, Mr. Hang, the sundry shop owner and colleague of Mr. Wu, handed a reporter photos, clippings, and documents collected over a decade of environmental work. He said he had no use for them now. Environmental work had become too risky.

He said he had recently seen some little fish darting around in the milky green water of a canal nearby. He took it as a good sign. "Once the white shrimp come back, that would be good," he said. The white shrimp had not come back just yet.

JOHN SEABROOK

Sowing for Apocalypse

FROM *THE NEW YORKER*

Deep inside a mountain on a tiny Norwegian island, seven hundred miles from the North Pole, lies the planet's ultimate safeguard against global famine: a vault housing a collection of seeds from around the world. John Seabrook ventures to this remote, desolate spot with the remarkable man who is the world's seed banker.

A COLD DRIZZLE WAS FALLING over St. Petersburg last March, and the gray morning light filtered through the grimy windows of the ceremonial rooms of the Vavilov Research Institute of Plant Industry, one of the oldest seed banks—and the most storied—in the world, situated on St. Isaac's Square. In one of the rooms, a woman in a smock sat at a table with a brown packet, and its contents, pea seeds, spilled out over the table in front of her. She did not look up from sorting through the seeds as two visitors passed, and, with her lips moving silently, she appeared to be lost in thought, or prayer.

Cary Fowler had an appointment to meet the director general of the Vavilov Institute, Nikolai Dzyubenko, in order to discuss the institute's seeds. Fowler, an American, is the world's seed banker. It's a nebulously defined position, yet a critical one. As the executive director of the Global Crop Diversity Trust, which funds the Svalbard Global Seed Vault, in Norway, Fowler is engaged in the Noah-like task of gathering the seeds of about two million varieties of food plants—both the familiar domesticated crops and many of their wild relatives—in order to create the first global seed bank.

We tend to imagine apocalypse coming in the form of a bomb, an asteroid, or a tsunami, but should a catastrophe strike one of the world's major crops Fowler and his fellow seed bankers may be all that stand between us and widespread starvation. Any of the diseases currently active in the United States—the rust fungus attacking soybeans; the potato late blight (the same one that caused the Irish potato famine), which turns potatoes into a black mass of rot; the Western bean cutworm, which feeds on corn plants—has the potential to turn into a devastating nationwide scourge. Should that happen, the only remedy—genetic resistance—might lie in an obscure variety, stored in a seed bank.

The Vavilov Institute is a monument to the extraordinary sacrifices people have made in order to save seeds. During the winter of 1941–42, when Hitler's troops were blockading Leningrad, cutting off food and supplies, the scientists who worked there protected the seeds stored inside the buildings, which amounted to several tons of nutritious food, from the starving Russians outside. At night, thousands of rats would invade the laboratories; the staff guarded the seed collections with metal rods. When some collections of potatoes needed resowing during the winter, institute workers found a plot outside Leningrad, near the front. Eventually, much of the collection was smuggled out over frozen Lake Ladoga, to a hiding place in the Ural Mountains. A. G. Stchukin, a specialist in peanuts, died of starvation in the building, as did D. S. Ivanov, a rice specialist, both surrounded by thousands of packets of seeds.

The story of what happened at the Vavilov Institute has a mythic resonance in the mind of every seed banker, and Fowler glanced around almost reverently as he walked through the shadowy halls. One of his personal heroes is Nikolai Vavilov, the Russian biologist and plant breeder for whom the institute is named—and the first man to dream of creating a world seed bank. For Fowler, coming here, to arrange for the institute to send seeds to the Svalbard vault, in time for its opening, in February 2008, was, in a sense, finishing the job that Vavilov had started. And Vavilov himself was following in a tradition of seed saving that reaches back into prehistory.

AGRICULTURE IS THOUGHT TO have begun around 8000 B.C., in the semi-arid mountains of Mesopotamia. Flint sickles and grinding stones discovered in the region suggest that the first farmers collected wild grains, which were developed over time into wheat and barley. Plants were also domesticated by other civilizations in other parts of the world, almost certainly independently. In Southeast Asia, farming began with the domestication of rice, around 6500 B.C.; in Mesoamerica, maize and squash were domesticated between 8000 and 5000 B.C. In each case, a legume was domesticated along with a grain or cereal: lentils with wheat in the Mediterranean; beans with maize in South America; soybeans with rice in Asia. Eating both together provided early humans with the right balance of protein and fat. Of the two hundred and fifty thousand known plant species in the world, only about two hundred are cultivated for food, and the vast majority of the world's food comes from just twenty crops, in eight plant families. It is a measure of the skill of the early farmers that almost all the plants we use in agriculture today were domesticated before historical times.

From the beginning, farmers must have realized that by saving a certain portion of the seeds from the previous year's crop they could insure themselves of a future harvest. (In Jarmo, Iraq, archeologists have found seed deposits that date from 6750 B.C.) Seed saving was one of the most important acts that a farming community performed. Seeds had to be protected from weather and animals—insects as well as mammals. One early method of preservation was to pack seeds and ash inside baskets, and then bury the baskets in the ground. Seeds were also sealed inside adobe structures, and kept in elevated thatched huts. When the community moved, it took its seeds along, too.

Biologically, a seed is an embryo of a plant. Around the embryo is usually a layer of endosperm, where the food for the embryo is stored, and around that is the seed coat, which protects the embryo until its moisture and heat sensors say that it is time to germinate. The embryo can survive for many years, under the right conditions, but not forever. In the 1890s, when the tombs of the Egyptian pharaohs were being opened by archeologists, hucksters tried to promote what were said to be ancient Egyptian wheat seeds—the idea being that after such a long rest the seeds would be especially productive. But there's no evidence that any of these seeds germinated.

A seed is also a plant's legs. Wind and water spread seeds, as do birds, bears, foxes, and many other animals, but man has proved to be the longest-distance distributor. When Columbus arrived in the New World in 1493, on his second voyage, he brought the seeds of plants known only in the Old World, among them wheat, onions, citrus, melons, radishes, olives, grapes, and sugarcane, and he took away seeds of plants known only in the New World, including corn, potatoes, tomatoes, peppers, pumpkins, squash, pineapples, and sweet potatoes. During the colonial period, the world's ecological boundaries were redrawn, as domesticated plants were carried far afield, and used to establish agricultural economies in other parts of the world. The Royal Botanic Gardens, at Kew, outside London, was the headquarters of Great Britain's botanical empire; from there, administrators coordinated the efforts of plant collectors at regional botanical stations from Jamaica to Fiji. As Lucile Brockway explains in her classic book *Science and Colonial Expansion*, published in 1979, cash crops, taken mostly from Latin America, where labor was scarce, were planted in Asia, where labor was abundant. The cinchona tree, from whose bark quinine is made, was transported from the Andes, and then planted in India by the British; the antimalarial treatment then enabled the colonization of Africa. The British also took natural rubber from Brazil, where the plant was first domesticated, and created a rubber industry in Southeast Asia that by the 1920s had greatly diminished Brazil's share of the rubber business. Sugarcane, which probably originated in India, went west, and became the main plantation crop of the West Indies. Coffee from Ethiopia was domesticated and introduced to India by the Arabs, and then cultivated by the Dutch in Java. Most of the coffee that grows in Latin America today traces its ancestry to a single coffee plant from Java that was taken to the Amsterdam Botanic Garden in 1706.

BEFORE SEEING DZYUBENKO, FOWLER was shown around some of the old rooms by a member of the institute, a tall, thin man with a long beard. Most of the exhibits concerned the life of Vavilov, who was born in 1887, the son of a prosperous Moscow merchant. Vavilov's education as a plant breeder coincided with the rediscovery, in 1900, of the work of Gregor Mendel, an Austrian monk who had died in 1884, and whose pea-breeding experiments were overlooked during his lifetime. Mendel established the fundamental laws of inheritance, and Vavilov, among others, was prescient enough to grasp their implications. Plant breeding, which had hitherto been an art, would now be a science. By crossing and backcrossing progeny with parents in order to isolate desirable qualities—higher yields, stronger roots, frost resistance—plant breeders could select traits from a broad spectrum of varieties, and combine them to create superior seeds. In order to realize the power of these new tools, however, breeders needed easy access to a large pool of genetic diversity. That was the quest to which Vavilov devoted his life. With the aim of creating hardier and higher-yielding Russian crops, Vavilov embarked on a series of expeditions to collect and catalogue ancient domesticated varieties (known as "landraces") of wheat, barley, peas, lentils, and other crops, as well as their wild relatives, reasoning that because they were well adapted to their natural environments they must contain valuable genes that could be incorporated into Russian crops. Over the next two decades, Vavilov himself collected more than sixty thousand samples, in sixty-four countries; altogether, his teams collected two hundred and fifty thousand samples. The present-day collection is based on those seeds.

Most people understand crop diversity in terms of choice—it's the difference between the sweet, creamy flavor of a Gala apple and the tart, crisp taste of a Granny Smith. But agricultural diversity is much more than that; it is a record of more than ten thousand years of human experience with crops, and of the struggle to produce food in changing ecosystems and climates. Crop diversity may be the most precious natural resource we have, because, as Stephen Smith, a research fellow at Pioneer Hi-Bred, one of the world's largest seed companies, has said, "How humans use diversity in farming determines our food, our health, and our economic well-being, and that in turn determines our political security." Crops such as corn and potatoes have been forced to adapt to vastly different climates in distant places: climate change has been a constant in the lives of crops for millennia. That is why seed banks, which are the primary repositories of crop diversity, are so important: the genes in them may represent our best hope for feeding ourselves in a warming world.

Vavilov observed that crop diversity is scattered unevenly around the world. There are certain places where it is abundant—Asia Minor for wheat and barley, the Andes for potatoes—and other places, such as Russia and the United States and Northern Europe, where there is very little. Vavilov eventually mapped eight centers of diversity, loosely grouped in a belt around the planet's middle. (Scientists speculate that the last Ice Age killed off diversity in much of the Northern Hemisphere, leaving a small center around the Mediterranean, where asparagus was first domesticated, and another in Eastern Europe, where barley and peas grew in abundance.) Building on the earlier work of a Swiss botanist named Alphonse de Candolle, Vavilov developed a theory that since diversity occurs over time, the centers of greatest diversity must also be the centers of origin for those crops. In 1926, he published *The Centers of Origin of Cultivated Plants.* Vavilov's insight (which was subsequently qualified by other researchers) became the basis for national claims of sovereignty over seeds.

But war and politics prevented Vavilov from realizing his dream of a world seed bank. Like so many other Soviet scientists, he fell afoul of Stalin. Vavilov came from a wealthy family, was not a member of the Communist Party, and was friendly with Nikolai Bukharin, a rival of Stalin's. Furthermore, genetics was considered a form of "metaphysics," and geneticists the enemies of Bolshevism. Vavilov was arrested in 1940, charged with treason and espionage, and interrogated, sometimes under torture, for eleven months. His trial was held on July 9, 1941; the tribunal took only five minutes to find him guilty and sentence him to death by firing squad. Later, Vavilov's sentence was commuted to twenty years in a prison at Saratov, on the Volga River. There Vavilov died, of starvation, on January 23, 1943. He was buried in a common grave.

Fowler was familiar with these details of Vavilov's life, but he listened politely as his guide repeated them. Fowler is six feet tall, fifty-seven years old, with curly reddish hair, glasses, and a Southern accent and courtliness that derive from his upbringing, in Memphis, Tennessee. He had not removed his green parka—the rooms were chilly—and he carried a briefcase, which contained artists' renderings of what the Svalbard vault will look like. Dressed in a blue blazer, with a buttoned shirt collar hanging loosely around his neck, he looked like a schoolboy who had spruced up for a class photograph. His eyes are deep-set, and his prominent forehead sometimes makes it hard to discern whether they are open or shut. This morning, he looked very pale; he had told me earlier, as we left the hotel, that he thought he was coming down with the flu. But that wasn't going to keep him from his appointment with the director.

In addition to discussing the transfer of some of the institute's seeds, Fowler wanted to broach the delicate subject of what condition the seeds were in. As a seed bank, the Vavilov has been in long decline. "No one I know would claim to know what is really happening there," Fowler had written to me before the trip. "The most important question being, To what extent are the collections still alive? Certainly much has died. But how much? Some experts in particular crops claim that a lot still exists. But for other crops we know that the conditions of conservation/regeneration could only have led to large losses."

Nikolai Dzyubenko occupies Vavilov's former office, on the south side of the square. The wooden floors creaked and groaned loudly, as we approached the room. Dzyubenko wore pink-tinted glasses, and his eyes were expressionless. He sat directly underneath a portrait of Vavilov, whose face was smiling, eager, and full of energy and optimism. The room looked both grand and tawdry; all the old elegance had floated up to the ceiling, like smoke, and clung to the elaborately painted plaster and the chandeliers. Fowler, gritting his teeth against his worsening flu, took out his papers and began to describe the Svalbard Global Seed Vault.

"It's seventy-eight degrees north, very remote, and the town has an excellent infrastructure," he said. "There will be a big ceremony in February, for the opening, and we'd very much like to work with you on getting some of the Vavilov seeds sent there in time for that." A heavyset man, who was wearing a green sweatshirt that said "Australia" in yellow letters on the front, interpreted.

Dzyubenko answered that, technically, moving the seeds to Svalbard would not be difficult, "but since we are not independent, and since the Vavilov collection is a public treasure, this must be discussed with our superiors at the Academy of Sciences in Moscow—the decision cannot be made at this level. So it will take some time to discuss it." I glanced over at Fowler, to see if this prospect—waiting for the wheels of Russian bureaucracy to turn, before seeds from the institute could get to Svalbard—discouraged him at all. But he merely pressed his lips together, and nodded.

After their talk, Dzyubenko flung open the doors to the next room, to reveal a small feast of pastries, fruits, cold meats, cheeses, juice, and vodka that had been prepared in the American's honor. Fowler felt too queasy to eat any of it, though he did manage to touch the vodka to his lips for a toast. I thought that he would skip the tour of the storage facilities that the director had planned, but he insisted on going through with all of it, including a visit to a new cold-storage room. Another room, full of large stainless-steel liquid-nitrogen storage vats, looked impressive—until Fowler noticed that the digital readouts on all but one of the vats said "Error."

IN THE BRONZE AGE, when agriculture had become firmly established as a primary source of food, few calamities would have been as devastating to a community as the loss of its seed stores, or the destruction of its crops. But in our time we almost never hear about these kinds of catastrophes. During the United States–led invasion of Iraq, in March 2003, the looting of Iraq's national archeological museum received considerable attention, but almost no one noted that the country's national seed bank was destroyed. The bank, in the town of Abu Ghraib, contained seeds of ancient varieties of wheat, lentils, chickpeas, and other crops that once grew in Mesopotamia. Fortunately, several Iraqi scientists had placed samples of the country's most important crops in a cardboard box and sent them to an international seed bank in Aleppo, Syria. There they sit, on a shelf in a cold room, waiting for a time when Iraq is stable enough to store them again.

Afghanistan's bank, which contained rare varieties of almonds and walnuts, and also fruits including grapes, melons, cherries, plums, apricots, peaches, and pears—many of which originated in the region—was destroyed in the 2001 overthrow of the Taliban. Scientists in Kabul had taken the extra precaution of hiding the national seeds in the basements of two houses in the towns of Ghazni and Jalalabad. But when they returned after the fall of the Taliban they discovered that looters had dumped the seeds on the floor. "Apparently, they were after the jars," Fowler told me. Those randomly scattered seeds represented dozens, perhaps hundreds, of unique varieties—Afghanistan's agricultural heritage.

Natural disasters can also destroy seed banks. Last year in the Philippines, a typhoon flooded a seed bank; there were reports of jars of seed floating in the ocean. In 1998, Hurricane Mitch demolished the national seed bank of Honduras. Nicaragua lost its national seed bank in the 1971 earthquake. Or banks can simply succumb to neglect.

Most of the fourteen hundred public and private seed banks in the world appear to be in a less precarious condition. There are national agriculture banks, which contain the seeds of crops grown in an individual country. There's also a network of international seed banks, funded by some sixty countries and organizations and managed by the Consultative Group on International Agricultural Research, which store specific crops. The bank in Aleppo, where the Iraqi scientists sent their seeds, has one of the world's largest collections of wheat and barley seeds. The main rice seed bank is in Los Baños, the Philippines, and the maize bank is in Mexico City. There are also banks for wildflowers, trees, and wild species of plants. Some are "ex situ"—off-site—and others are "in situ," conserved in fields. The New England Wild Flower Society preserves North American native plants in "sanctuaries" throughout the Northeast. In 2000, the British Royal Botanic Gardens launched the ex-situ Millennium Seed Bank, an ambitious project (and a favorite of Prince Charles) that includes preserving all the native seed-bearing species growing in Great Britain. There is a movement to construct banks for disappearing breeds of domesticated animals; sperm and embryos are cryogenically preserved in liquid nitrogen, at a temperature of minus one hundred and ninety-six degrees Celsius. However, when it comes to the wild species of the world, animals as well as plants, our current preservation efforts are grossly inadequate. At the recent G8 summit in Germany, in June, scientists predicted that as much as two-thirds of the world's wild species could be nearly extinct by 2100, because of habitat destruction, overfishing, and climate change. The resulting explosion of pests and the loss of pollinators would be only two of the devastating consequences for agriculture.

And even a well-run bank isn't an iron-clad guarantee against extinction. Most seed banks were created as short-term storage facilities in order to develop new seeds—something like genetic libraries. In the past thirty years, they have been modified to include long-term storage facilities, which preserve varieties that are no longer grown in fields—they're more like museums or zoos. Many gene banks aren't adequately equipped or funded for long-term storage. Fowler told me, "We think that 50 percent of the unique collections in developing nations are in danger. Half. That's pretty stunning, when you think about it. As one scientist said to me, 'We call them seed banks, but actually they're more like morgues.'"

FROM THE BEGINNING OF the United States' history, its people have been preoccupied with seeds. The early settlers faced a landscape largely devoid of domesticated crops, with the notable exception of maize, which Native Americans had brought from Central America. Among economic crops, only blueberries, cranberries, hops, and a type of sunflower originated in North America; a meal made exclusively of local ingredients would be meagre. Therefore it was necessary to import plants and seeds from other countries. Thomas Jefferson, who once wrote, "The greatest service which can be rendered to any country is to add a useful plant to its culture," smuggled rice seeds out of Italy by sewing them into the lining of his coat. Just as immigration brought cultural diversity to the United States, so the immigrants brought botanical diversity, in the seeds they carried with them, which were often concealed in the brims of their hats and the hems of their dresses. In 1862, in the midst of the Civil War, Congress, at the urging of Abraham Lincoln, established the Department of Agriculture, in order to collect "new and valuable seeds and plants . . . and to distribute them among agriculturalists." Great plant explorers like David Fairchild, who was Alexander Graham Bell's son-in-law, and Frank Meyer, for whom the Meyer lemon is named, introduced new crops to the United States, where they thrived.

By 1898, when the USDA established the Office of Foreign Seed and Plant Introduction, the government was distributing some twenty million seed packages a year to farmers. A network of state breeding stations helped develop the most productive varieties for each region. Beginning in the late 1940s, the government established regional seed banks that focused on individual crops: a center in Ames, Iowa, was devoted to maize; the apple and grape research station was in Geneva, New York; the potato center was established in Sturgeon Bay, Wisconsin. In the 1950s, the government constructed a national seed bank—the Fort Knox of seeds—in Fort Collins, Colorado, the cornerstone of what is known today at the National Plant Germplasm System. The bank, which I visited in December of last year, holds nearly five hundred thousand kinds of seeds—its holdings include both varieties grown domestically and backups for other, international collections. It is a model for the Svalbard vault. The main storage vault is kept at eighteen degrees below zero Celsius: the ink in my pen froze as soon as I entered the room.

Fowler explained the basic principles of storing seeds in banks. When the seed comes in from the field, it is sorted, labeled, cleaned, and dried to a humidity level of about 5 percent. Dryness and cold are the most important factors, to slow down the seed's metabolism, and to insure that it won't germinate. Breeders' collections are generally stored at room temperature or a little below and are intended to last for only a few years; base collections are kept at between minus ten and minus twenty degrees Celsius, a temperature at which some seeds can live for more than a hundred years. (Grains, such as wheat and barley, tend to live the longest.)

In any well-run bank, samples of the seeds have to be regularly germinated, to insure that the seeds are still viable. If the germination rate drops below a certain point, new plants must be grown in the field from the seed and new seeds collected from those plants. "It's not too complicated, but there's a lot of labor involved, and it's expensive," Fowler said. "Plus, there are equipment failures, poor management, funding cuts, natural disasters, civil strife—you name it."

Fowler's interest in agriculture began on his maternal grandmother's three-hundred-acre farm, near Madison, Tennessee, which he visited every summer as a boy. "There was cotton, corn, soybeans, chickens running in the back yard, a couple of milk cows—it was a real old-fashioned farm," he told me. "We'd go to the experimental plots at the local agricultural station for our seeds, and my grandmother would talk to everyone about the different varieties and make selections. She wanted me to take the farm over. She would always ask me, 'Do you want to be a farmer?' We spent a lot of time driving down dusty roads while she gave a running commentary on the quality of crops and soils, pointing things out. But I was more interested in the stuff I was studying in school—Sartre, freedom and determinism, the role of the individual in society, that kind of stuff. And I knew I wanted to be politically active, though I wasn't sure in what area."

During the school year, Fowler lived with his parents, in Memphis, where his father was a defense attorney and, later, a judge. He took an active part in civil-rights demonstrations, and was present in the church on the night that Martin Luther King made his last speech; King was assassinated in Memphis the next day. When he graduated from Simon Fraser University, in Canada, in 1971, he received conscientious-objector status, "which greatly upset my father, who had enlisted the day after Pearl Harbor," Fowler told me. After working in a hospital in North Carolina, as a clerk, for about a year, he was released from service, and planned to go to Sweden, to study for a Ph.D. in sociology at Uppsala University, forty miles north of Stockholm.

That year, however, Fowler discovered an oddly shaped mole on his stomach. By the time it was biopsied, the cancer had spread all over his body.

" 'Do you have life insurance?' the doctor asked me," Fowler recalled.

" 'No,' " I said. " 'Why?' "

" 'Because you have six months left to live.' "

The doctors did what they could, removing part of his stomach where the melanoma had appeared, but the prognosis was still grim.

"So I went home and waited to die," Fowler said. "Every time I'd feel the slightest twinge in my body, I'd wonder, Is this the cancer? Is it starting? Am I going to die now?" But he didn't die, and after about a year he decided, "This is no way to live," and he went back to pursuing his sociology Ph.D. The doctors were so astounded that they sent Fowler to Memorial Sloan-Kettering, in New York, for a test of his immune system. Although it seemed to weaken when it was resisting the cancer, it became remarkably effective once the cancer started.

"Ten years later, I was given a diagnosis of testicular cancer," he said. He underwent a painful procedure that involved injecting dye into his feet and then circulating it through his lymphatic system; an X-ray would show how far the cancer had spread. "I lit up like a Christmas tree," Fowler said. "The cancer was everywhere. Though they didn't tell me at the time, they had never had a patient who had survived that kind of cancer." The doctors elected to do extensive radiation treatments; Fowler still has a map of tattoos all over his body that guided the radiation machine. Once again, Fowler told me, "The cancer just disappeared."

I asked how his cancers had influenced his work in saving seeds. Fowler replied, "The first one, I didn't handle it very gracefully. I was scared. Really scared. And the reason I was scared was that I hadn't done anything—I hadn't contributed constructively to society. And that was frightening."

FARMERS BEGAN TO TURN away from the ancient practice of saving seeds early in the twentieth century. Plant breeders had discovered that, when two inbred lines are crossed with each other, the next generation explodes in "hybrid vigor," producing more robust plants than those which were allowed to pollinate randomly (known as "open pollination"). If the progeny of two pairs of inbred lines are themselves crossed (a "double cross"), their offspring will be even more vigorous. However, if those crosses and double crosses are then allowed to reproduce naturally, through open pollination, only a fraction of their progeny will show hybrid vigor.

Corn is among the easiest plants to hybridize, because the male parts, which are the tassels that contain the pollen, and the female parts, which are the ear and the silks, are widely separated. It is relatively simple, though labor-intensive, to cross two inbreds by sowing the two lines side by side in nearby rows, and removing male parts of the plants from one line, to insure that it is fertilized by the other. (This method of emasculating corn plants still provides summer employment to thousands of teenagers throughout the Midwest.) American agriculturalists in general were slow to recognize the potential of hybrid corn, but several American breeders championed the new technique. One of them, Henry A. Wallace, happened to be the son of Warren Harding's Secretary of Agriculture, Henry C. Wallace, and his enthusiasm was heard in high places. In 1924, a Henry A. hybrid, which he dubbed Copper Cross, won a gold medal in the Corn Show at the Iowa Corn Yield Test. In 1926, Wallace founded the Hi-Bred Corn Company, later Pioneer Hi-Bred, to market his seeds.

Pioneer was by no means the first private seed company. A seed trade, catering to farmers who didn't want to take the time to clean and sort their seed, had existed since the early 1800s. But farmers had only to buy the seed once, and then generate more seed themselves. With hybrids, however, farmers had to buy the seed every year if they wanted to enjoy the benefits of hybrid vigor. From a commercial point of view, plant breeders had hit the jackpot—a "biological patent" on seed. Pioneer reaped the benefits of this good fortune, and eventually became the dominant seed company in the world.

In 1933, hybrid corn amounted to about one half of 1 percent of the planted corn acreage in the United States. By 1945, thanks in part to promotion by the USDA, that figure had risen to 90 percent. Throughout the Depression, American farmers, who could have grown and saved their own seed by using traditional open-pollinated varieties, instead bought hybrid seed from corn companies; the increased yields justified the expense.

Beginning in the 1940s, American-made hybrid seeds were sent around the world, as part of a vast agronomic program that came to be known as the green revolution. Norman Borlaug, an American plant breeder, used a strain of Japanese semi-dwarf wheat, known as Norin 10, which had been bred in Japan and brought to the United States in 1946, during the Allied occupation, to create a wheat with a stalk short enough to support a larger, more productive head. First in Mexico, and later in Pakistan and India, Borlaug's wheat allowed local farmers to double, and in some cases quadruple, their yields. In 1966, the International Rice Research Institute created a variety of stunted rice called IR8, a cross of an Indonesian type with a Chinese strain, which was widely planted in Asia.

The green revolution was a complicated blend of altruistic and imperial motives, played out through seeds. The notion that humans now had the power to banish the spectre of starvation and famine, which has haunted our species for millennia, was a potent one. The green revolution is estimated to have fed roughly a billion people who might otherwise have starved. In developing countries, production of cereals doubled. By the mid-1980s, the average person in these countries consumed 25 percent more calories per day than in the early '60s. But the development and distribution of the super-seeds, which was funded by the World Bank, the United States seed trade, the Rockefeller Foundation, and the Ford Foundation, was also a clever way of planting American-style agrarian capitalism in developing nations that might otherwise be in danger of succumbing to Communism. In Fowler's 1993 book *Shattering,* written with Pat Mooney, a Canadian activist, he points out that the new hybrids "produced not just crops, but replicas of the agricultural systems that produced them. They came as a package deal and part of the package was a major change in traditional cultures, values, and power relationships both within villages and between them and the outside world." Now, instead of growing crops for local consumption, farmers began growing crops for export. And, like the American farmers before them, Mexican, African, and Asian farmers lost the incentive for saving seed.

In 1973, Fowler started working for a journal called *Southern Exposure.* "It was dedicated to improving the image of the South, and they were working on an issue about the disappearance of family farms," he said. "I got really involved in the subject. I had grown up in these two worlds, with a love of agriculture but with no sense of how that would fit in with what I was really interested in—politics, the law, social justice. Now I began to put things together." Two years later, he became a researcher for Frances Moore Lappé, who was writing a book called *Food First.* Lappé was the author of *Diet for a Small Planet,* which became a bestseller in 1971, promulgating the message that Americans could help solve world hunger by shifting to a predominantly vegetarian diet. But, with *Food First,* Lappé wanted to analyze global food policy. Like many others, she was beginning to see that the aims of the green revolution weren't as simple as they appeared. Hunger was not only the result of a scarcity of food; it could also be caused by a production system that replaced traditional, sustainable agriculture, as practiced by peasant farmers, with a global export system that was driven by foreign agribusinesses. Lappé had rented a house in Hastings-on-Hudson, where Fowler spent a year, working with Lappé and her coauthor, Joseph Collins.

In the course of his research, Fowler read Jack Harlan's articles on the loss of crop diversity, including "The Genetics of Disaster" and "Our Vanishing Crop Genetic Resources." Harlan, a professor of genetics at the University of Illinois, argued forcefully that the adoption of modern hybrid seeds around the world was causing the traditional varieties, grown by farmers for millennia, to become extinct. Landraces were, after all, thoroughly domesticated; if they weren't cultivated, they couldn't survive. In their place, the hybrids created monocultures. And, because hybrids are created by crossing purebred lines, these monocultures contained a narrower spectrum of genes. That meant that a single disease could wipe out much of the national crop. In the spring of 1970, a type of corn blight invaded cornfields in the southern United States. By the end of the year, it had killed 15 percent of the American crop; some southern states lost 50 percent of their corn. In 1972, the National Academy of Sciences released a report on the genetic vulnerability of major crops, which found that 70 percent of the United States corn crop consisted of just six varieties.

The very success of plant breeders' efforts was eliminating the raw material that made their work possible. In the United States, the nation's agricultural diversity, which had been rich in 1900, was vanishing from fields. A survey in 1983 found that, since 1903, the number of readily available varieties of cabbage dropped from 544 to 28; carrots dropped from 287 to 21; cauliflower varieties fell from 158 to 9; and varieties of pears fell from 2,683 to 326. In many cases, the new commercial hybrids that replaced the traditional varieties no longer tasted as good—they were bred more for production than for flavor.

Farmers, enjoying vastly greater yields with the new hybrids, couldn't be expected to go back to planting the traditional varieties. (The backlash to industrial monocultures did, however, help to inspire widespread interest in "heirloom" seeds, which began with the founding of the Seed Savers Exchange, in 1975, and also led to the creation of local farmers' markets where at least some of the old varieties can be found.) In the case of many landraces, the only alternative to extinction was preservation in the breeders' ex-situ storage centers—the seed banks. But most of these banks were short- or medium-term storage facilities, and even the long-term centers, like the national seed bank at Fort Collins, were at the time poorly funded and staffed, and hardly qualified to serve as the last line of defense against the mass extinction of landraces.

Fowler's undergraduate thesis had focussed on Jean-Paul Sartre's notion of human agency in politics, and he wanted to concentrate on the issue of genetic-resource preservation. In 1977, Fowler began working with Pat Mooney, who persuaded him to direct his political activism at the principal international venue within which food-policy matters are decided—the United Nations Food and Agriculture Organization, or FAO, which is based in Rome. Together, they marshaled opposition on the issue of patenting of seeds, which had come to the fore with the passage of the United States Plant Variety Protection Act, in 1970. This law, which was strengthened in 1980, and supplemented that year by a Supreme Court decision allowing the patenting of novel forms of bacteria, gave plant breeders a broad legal basis for claiming ownership of genetic resources.

As far back as Luther Burbank, the celebrated American plant breeder who was a contemporary and friend of Thomas Edison, breeders had complained about the injustice of awarding intellectual-property protection to novel mechanical devices and denying it to botanical inventions, even though a new plant might benefit millions. Soybeans, for example, which constitute the fourth-largest crop in the world, are self-pollinated, and can't be crossbred easily. Before the Plant Variety Protection Act, a breeder who wished to develop an improved variety of soybean could not expect much of a return on his investment, because farmers would have to buy the seed only once. But, with the protection of the act, which made it illegal to save patented seeds, the seed industry could justify much greater investment in research and development, because farmers would have to buy new seed or pay a license fee every year.

From the perspective of many developing nations, however, the American-led movement to patent seeds was an outrage. The industrialized countries of the North, having helped themselves to genes from the centers of diversity in the South, and having used them to create agricultural industries, were now asking the South to buy those genes back, in the form of patented seeds. According to Jack Kloppenburg, a professor of rural sociology at the University of Wisconsin, crops that originated in the Near East and Latin America make up 66 percent of global food crop production; crops originating in North America and Europe represent less than 5 percent, combined. The North was guilty of "bio-piracy"—a slogan that became popular in the 1980s. Armed with Vavilov's theory that the centers of diversity must be the centers of origin, India, Brazil, and Iran, among other countries, began to press their case in the FAO for sovereignty over these resources. If the North wanted to claim ownership of patented seeds, the South would claim ownership of the genes from which those seeds were made. The Seed War was engaged.

By the mid-'80s, a "genetic OPEC" had begun to take shape. On coffee plantations in Central America, where a disease called coffee rust was a common threat, breeders wanted to return to Ethiopia, the origin of the coffee plant, and find a variety within the seed bank there that would be rust-resistant. But Ethiopia wouldn't allow breeders access to the coffee genes stored in its seed bank. The Jamaican government refused to allow the genes of allspice to be exported; India did the same with black pepper and turmeric seeds; Ecuador locked up its cacao seeds; Taiwan embargoed sugarcane; and Iran denied the world access to its pistachio collection.

Back in 1979, Fowler and Mooney had begun attending the FAO's annual conferences in Rome. The FAO, which was heavily weighted with Third World representatives, offered a mostly sympathetic forum to the notion of seeds as a genetic commons, part of the shared heritage of humanity, and it was generally hostile to the notion of seeds as commodities. At the 1981 conference, the Mexican delegation, having consulted with Fowler and Mooney, proposed the idea of an international seed bank. It would contain seeds from national and international seed banks, and patented seeds created by private seed companies. The North could have free access to the seeds from the centers of diversity only if the South could have free access to the patented seeds. This proposal was enthusiastically embraced by many developing nations, and, at the 1983 conference, the international seed bank became a central part of a nonlegally binding document known as the International Undertaking on Plant Genetic Resources for Food and Agriculture, which was signed by more than 110 countries around the world.

The bank was denounced by seed companies in the United States and other developed countries. The American Seed Trade Association said that the Undertaking "strikes at the heart of free enterprise and intellectual property rights." Seedsmen argued that patented seeds represented an enormous investment of labor and capital on their part; the unimproved coffee seeds in Ethiopia's seed bank weren't valuable until that investment was made. But Fowler maintained that even primitive landraces represented many generations of selection by farmers who lived around the centers of diversity, going back thousands of years. As he wrote in his book *Unnatural Selection*, which was published in 1994, "Can we really say, that the modern plant breeder who turns out a disease-resistant tomato, wheat, or rice variety has done something more grand or worthy of reward than the farming community that first identified and conserved the disease-resistant characteristic in its fields?" He and Mooney began to advocate the concept of "farmers' rights," demanding a monetary system that would recognize the contributions made by farmers in developing nations from which the seeds were taken. As Fowler explained to me, "We were pointing out how unfair it was that after a century and a half of free and unimpeded gene flow from South to North, for the North to come along and say, OK, now you have to pay for the genes—genes we took from you—because now we have a patent on them."

GENE-SPLICING, FIRST PERFORMED BY Stanley Cohen and Herbert Boyer in 1973, allowed seed companies to offer a limitless range of products. Today's plant breeders can incorporate genes from plants that are not sexually compatible with each other. Indeed, because all living things share the same coded language of DNA, breeders can choose genes from outside the plant kingdom altogether—genes from bacteria, and even from fish, were used to create new kinds of genetically modified organisms, or GMOs. This technology, in turn, led to a new type of seed company, the avatar of which is Monsanto, an agrichemical corporation headquartered in St. Louis, which is today the largest seed company in the world; other leading seed corporations include Bayer, Syngenta, and Dow, all of which have roots in the chemical or the pharmaceutical sector, rather than in the seed trade. Whereas Pioneer was a seed company that turned to genetic engineering as a way of improving its product—seeds—Monsanto was a chemical company that saw seeds as a delivery vehicle for its product, which was genes. Instead of creating seeds through hybridization, as Pioneer did, Monsanto would license genetically engineered traits to seed companies. Monsanto's Bt corn, which was introduced in the mid-'90s, incorporated genes taken from a soil bacterium, *Bacillus thuringiensis*, that provides these Bt crops with resistance to certain pests. Last year, about 40 percent of the U.S. corn crop was Bt corn. Food made with this and other types of genetically modified organisms is sold in virtually every supermarket in the United States, and does not require labeling. However, in Europe, and in some other parts of the world, GMOs have attracted widespread protest by consumer groups, and are subject to much tougher regulations.

The consolidation of the seed industry, which began in the 1970s, has continued in the era of genetic engineering; 55 percent of the seeds used to grow the world's food are sold by just ten global corporations. Large amounts of capital are required to create today's superseeds: biotech companies estimate that the research and regulatory costs of getting a single new genetically engineered trait on the market are as much as a hundred million dollars; a seed that offers two or three "stacked" traits—the state of the art in seed technology—may cost three hundred million dollars. These new seeds could help us cope with climate change and population growth, by producing crops designed to survive on less water, in hotter conditions, with even greater yields. At Pioneer Hi-Bred, which was bought by DuPont in 1999, three-quarters of the company's seeds are GMOs; corn plants are being developed for increased ethanol production, drought tolerance, and resistance to rootworm, corn borer, and fungal disease. Many new plant varieties are being designed for industrial uses. Corn, soybeans, sunflowers, and canola can all be used in the production of ethanol and biodiesel. Soybeans are being developed to produce higher yields of oil for ink, with an aim toward replacing traditional petroleum-based inks used in printing. In the United States, corn yields continue to rise; last year, American farmers produced nearly eleven billion bushels, with an average yield of 149 bushels an acre. Worldwide, a new green revolution could be in the offing.

The bad news is that the rest of the world isn't as enthusiastic about the new superseeds, perhaps because some countries are still coping with the social consequences of the last green revolution. Other countries worry about the long-term effects of GMOs, which are banned from some markets. Many Europeans feel that genetic engineering is still too new and untested, and too many things could go wrong, from the unwanted pollination of conventional plants by the GMO plants to unexpected health and dietary effects caused by eating GMO food to an unforeseen blight that could wipe out a GMO monoculture. Moreover, the old idea of feeding the world's hungry masses, which helped sell the first green revolution, hasn't been embraced as warmly this time. Instead, there is suspicion among developing nations that such talk is all an elaborate "double cross." As Jack Kloppenburg wrote in 2004, in his book *First the Seed*, "The powerful tools of biotechnology are now being wielded largely by a narrow set of corporations which claim to want to use them to eliminate hunger, protect the environment, and cure disease, but which in fact simply want to use them as quickly as they can to make money just as fast as possible."

PLANT BREEDERS AND SEEDSMEN bitterly criticized Fowler and Mooney for politicizing the debate about genetic resources. By indirectly encouraging developing nations to deny plant breeders access to their seed banks, the activists were placing the responsibility for the conservation of those resources with countries whose unstable or dysfunctional governments couldn't be counted on to take the necessary measures. Many scientists felt that politics had no place in decisions about genetic conservation. Few wanted the governance of agricultural plants to follow the route that was eventually taken to guide the trade in medicinal plants—the Convention on Biological Diversity, which took effect in 1993. With the CBD, nations like Brazil, whose rain forests have long drawn bio-prospectors looking for new plant-based drugs and therapies, affirmed sovereignty over the genes within their borders. That system makes some sense when applied to the production of pharmaceuticals, which are distributed to narrowly focussed markets, but not to food, which is a worldwide resource. Even the countries within the centers of diversity depend heavily on imported crops to feed their populations. A similar treaty for world agriculture might mean that, in order to develop a variety of wheat using strains from numerous countries, individual agreements would have to be negotiated with every country. That was unthinkable.

At an FAO meeting that Fowler attended in 1981, the octogenarian Sir Otto Frankel, who was then among the world's most prominent plant scientists, angrily denounced Fowler in a Roman restaurant. Fowler told me, "Here I thought I was trying to do something good, and so to have one of the leading lights in the field just go off on me really made me question what I was doing." Among the people who witnessed the confrontation was Jack Harlan, who had inspired Fowler to take up the cause of genetic resources in the first place. Later, after Frankel left, Fowler went on, "I asked Harlan if he thought we were doing something that was bad, unproductive, destructive, and he said, 'I would never tell you that,' and then he said, 'They're going to fight you and call you names for five years, then there will be a period when things get a little bit better, and then in ten to fifteen years they will adopt all of your ideas but claim the credit for it themselves.' And that's pretty much what happened. At the third International Technical Conference, in 1983, I was viewed as a dangerous radical, and by the fourth conference, in 1996, I was in charge."

IN 1993, THE FAO HIRED Fowler to oversee the drafting of a Global Plan of Action for the Conservation and Sustainable Utilization of Plant Genetic Resources—a first step toward a rational, worldwide seed-bank management plan. The International Treaty on Plant Genetic Resources for Food and Agriculture, which was adopted by the UN in 2001 and has been ratified by most of the countries of the world, laid the legal groundwork for a global seed bank. The treaty is both a victory and a defeat for Fowler. It recognizes a version of the concept of "farmers' rights," and establishes a system of compensation for genes used in the creation of patented seeds. However, in return, the developing nations agree to drop their opposition to the principle of patenting seeds. In effect, Fowler has compromised one of his original principles, common heritage, in order to achieve the other.

Fowler and his colleagues knew where they wanted the vault to be—on the Norwegian archipelago of Svalbard. Although Norway took sovereignty over Svalbard in 1925, the archipelago has a long tradition as an international *terra nullius*—a no man's land. In the early 1980s, a backup site for the Nordic Gene Bank was created in a coal mine near the town of Longyearbyen. It contained the seeds from the five Nordic countries—Iceland, Finland, Norway, Sweden, and Denmark. Norway had proposed expanding the facility into a world seed vault, but, with the Seed War then raging, the idea was untenable. Now, however, with the treaty in place, and with a rising awareness of the ecological threat posed by climate change, the idea of an ultimate backup to all the other seed banks—a Doomsday vault—seized the imagination of officials in both Norway and Rome. In June 2006, the groundbreaking ceremony at Svalbard was widely covered in the press all over the world.

The vault is not the seed bank that Fowler envisaged back in the '80s—a common bowl from which the world could feed as one. Far from it: each nation will have access to only its own seeds. Nor does the vault much resemble the world bank that Vavilov envisioned—a kind of breeders' utopia, in which the seeds of every kind of food plant in the world are available. In 2006, 30 million new acres of GMOs were planted, bringing the total worldwide acreage to 252 million acres—about 7 percent of the world's cropland. Altogether, more than ten million farmers from twenty-two countries planted biotech crops in 2006. But few, if any, of these seeds are going into the Svalbard vault. Seed companies have shown little interest in putting them there, and, in any case, Norway severely restricts bringing GMOs into the country, even though, as Fowler put it to me, the seeds in the vault would be kept in "multi-ply packages, inside sealed boxes, inside an air-lock chamber, behind multiple locked doors, inside a mountain, frozen to minus twenty, in an Arctic environment in which no seed would survive even if it escaped."

As the use of biotech crops continues to grow, the seeds in the vault will represent an ever-smaller share of the seeds actually growing in fields around the world. An apt symbol of this is what happened in Iraq, the birthplace of agriculture, following the United States–led invasion. Even though Iraq's traditional varieties were preserved in Aleppo, the United States encouraged the use of seeds provided by American companies, many of them GMOs, which were distributed by the United States military as part of Operation Amber Waves. Order 81, issued in 2004 by Paul Bremer, the head of the Coalition Provisional Authority, prohibited Iraqi farmers from reusing these seeds, forcing them instead to purchase licenses from corporations to receive new seed each year.

I ACCOMPANIED FOWLER TO SVALBARD, to visit the site of the Doomsday vault. We flew from Oslo to Longyearbyen, a settlement of some two thousand inhabitants, and the northernmost destination on earth serviced by regularly scheduled flights; it's about eight hundred miles from the North Pole. From the air, I could see the craggy, snow-covered peaks thrusting some five thousand feet into the air—the mountain range discovered by the Dutch explorer Willem Barents in 1596, when he ran into it while trying to sail over the top of the world. For the next two hundred years, Svalbard was an international center of whaling, but by the end of the eighteenth century the bowhead whale had been hunted to extinction in these waters. In the eighteenth and nineteenth centuries, Svalbard was used by hunters, mainly Russians and Norwegians, whose greatest prize was the thick white coats of the polar bears that roamed the islands and the surrounding ice floes. (The polar bears survived, barely, although now climate change is threatening them with extinction once again.) The huts of some of the most famous hunters and trappers of the era remain, scattered on the snow.

Longyearbyen is on Spitsbergen, the largest of the islands, which means "jagged mountains" in Dutch. Its regular inhabitants live in neat wooden structures that are painted in bright colors. In the early twentieth century, Svalbard became a center of coal mining, and, on the crags above the town, the old mining trellises, which were used to bring coal out of the mines, can be seen: stark structures with thick cables strung between them. There is still some mining on Spitsbergen, but these days the Longyearbyen economy is driven mainly by the thousands of tourists who come during the summer months to explore one of the last true wildernesses in Europe, and to watch the Fourteenth of July glacier, at Krossfjord, on the northern side of Spitsbergen, calve into the Barents Sea.

Rune Bergstrom, the chief environmental officer for Svalbard, drove us to the site of the vault, halfway up one of the mountains visible from Longyearbyen. As we left town, he pointed out the graves of miners who died in the influenza epidemic of 1918. Scientists hoping to study the virus had taken tissue samples from the bodies in 1998. We followed a winding road up a steep ridge that was lightly covered with a fine, dry snow—a kind of Arctic sand. Along the road we saw two Svalbard reindeer: odd, goat-size animals, with comically stunted legs and long, awkward-looking antlers.

The vault is tunneled into the sandstone rock face of one of the mountains, near the old mine in which the backup to the Nordic Gene Bank is stored. The design is straightforward: from the entrance, a long tube lined with steel and concrete leads straight back to three large rectangular concrete rooms, which will house the seeds, on shelves. Surrounding the entrance of the shaft will be a twenty-seven-foot-tall concrete structure. A series of colored lights will be embedded in the top—an artist's installation. In November, as the four-month-long polar night settles over Svalbard, the lights will begin to pulse, producing a curtain of light that will change according to the lighting conditions of the Arctic. It will be visible far across the Barents Sea, out on the ice floes that the polar bears roam. When the sun returns, in late February, the lights will go out and the reflective surface will glow with the light of the never-setting sun.

To get a better sense of what it will feel like to be deep inside the mountain, we visited the Nordic bank. It was a relatively warm day—about twenty degrees Fahrenheit. In a mining shed, we donned blue overalls, helmets, headlamps, and gas masks, and followed a miner down the abandoned mine, stepping carefully over railway trestles on the tunnel floor. From time to time, we could see black seams of unmined coal in the walls. About two hundred yards in, we came to a large wooden door, which was frozen shut. The miner pounded on it with a sledgehammer until finally it swung open to reveal a smaller door, which was also frozen. Behind that was a large black steel cage, filled with crates. Frost covered the bars, and the ice crystals glistened in the beam of our headlamps. Fowler climbed inside the cage and opened one of the wooden lids. Inside, packed in foam peanuts, were sealed glass ampules, with acquisition numbers etched in the frosty glass. The ampules contained seeds, about five hundred of each variety, of 237 species.

Back outside, I wondered what would happen if the sea levels rise as much as some scientists predict they will. "We are a hundred and thirty meters above sea level," Fowler replied, his breath frosting. "The max sea-level rise under the worst-case climate-change scenario is eighty meters, so, whatever happens, the seeds should be safe." But who can predict what will happen if the ice melts? A northern sea lane could open year-round between Europe and Asia (the passage that Willem Barents was looking for in 1596, when his vessel became trapped in the ice), and Svalbard could take on the strategic importance that Malta held for centuries, when it was a midpoint between the Christian and Islamic worlds. Will the seeds themselves be viable in a world that warm? Even if they can grow, they won't have evolved the defenses necessary to ward off all the new pests and diseases that will appear. On the other hand, perhaps a few of the seeds inside the vault will hold the answers for the farmers of the future. "When you think about it, the plants have already been there," Fowler said. "When Columbus brought maize to Europe— that was a climate change. When maize then went to Africa, that was a climate change. We need to figure out how the plants were able to adapt to these changes, and repackage those traits."

Far away in the distance, I could see one of the old hunters' shacks, and I thought of the bowhead whale, hunted to extinction by eighteenth-century whalers, and of the polar bears, struggling for survival on the ice floes. Species extinction seems to be the baseline in humanity's relationship with the natural world; the notion of sanctuary is a relatively new and tenuous idea. But Fowler's craggy face, seen against the mountains, showed no trace of doubt.

About the Contributors

DANIEL CARLAT, M.D., is a psychiatrist in private practice in Newburyport, Massachusetts, and Assistant Clinical Professor of Psychiatry at Tufts University School of Medicine. He has been an occasional contributor to the Op-Ed section of the *New York Times*, and has written for both the *New York Times Magazine* and *Wired* magazine. He wrote a popular psychiatric textbook, *The Psychiatric Interview: A Practical Guide*, and he is the publisher and editor in chief of *The Carlat Psychiatry Report*, a monthly newsletter read by psychiatrists and other mental health practitioners both in the United States and abroad. He is involved in the American Psychiatric Association, where he sits on two committees examining the proper relationship between the pharmaceutical industry and the profession.

"Readers have wondered why I waited so long to write 'Dr. Drug Rep,'" he explains. "After the events recounted in the article (which occurred in 2002), I quit Wyeth's speaker's bureau and founded *The Carlat Psychiatry Report*, which receives no industry support. Over the ensuing years, the vast scope of the pharmaceutical industry's control of medical information became increasingly clear to me. As an editor, it is now difficult to find even a single national medical expert who doesn't require an asterisk next to his or her name because of a conflict of interest. I saw this as being an under-recognized problem and felt that describing in detail my own past involvement with the industry would help to awaken the public's indignation."

THOMAS GOETZ is the deputy editor of *Wired* magazine, where he acts as the magazine's trend spotter and oversees editorial projects. He also writes for the magazine. Most of his stories explore the confluence of medicine and technology, covering topics such as how computer models can forecast epidemics and the controversial diagnosis of metabolic syndrome. He holds a Master of Public Health degree from the University of California, Berkeley, and a Master of Arts in English from the University of Virginia. He is author of the blog Epidemix.org.

"The arrival of retail genomics has taken the science world by great surprise," he writes. "In the months that followed the debut of 23andMe, Navigenics, and other personal genomics start-ups, journals like *Nature* and the *New England Journal of Medicine* have reacted with some surprise and much caution, warning that the science isn't ready for consumer use. But there's no putting the genome genie back in the bottle—whether the status quo is on board, people want access to their DNA, even if the utility of seeing one's SNPs is still sketchy.

"As I see it, genomics is just the first step in a new health care paradigm, where informed individuals are routing around the medical establishment and using whatever data they can get—screening tests, their own medical records—in order to optimize their health.

"Whether the trend will result in a healthier population or a more hypochondriac one remains to be determined."

WINNER, WITH THE INTERGOVERNMENTAL Panel on Climate Change, of the 2007 Nobel Peace Prize, former vice president AL GORE is chairman of the Alliance for Climate Protection, a nonpartisan, nonprofit organization devoted to mobilizing global support for urgent and sustainable solutions for the climate crisis. He is cofounder and executive chairman of Current TV, an Emmy Award–winning, independently owned cable and satellite television nonfiction network for young people based on viewer-created content and citizen journalism. Gore is cofounder and chairman of Generation Investment Management, a London-based firm that is focused on a new approach to sustainable investing; and recently became a partner in the venture capital firm Kleiner Perkins Caufield & Byers, which has formed a strategic alliance with Generation to focus on solutions to the climate crisis. In addition, Gore serves on the board of directors of Apple and as senior adviser to Google.

Al Gore was elected to the U.S. House of Representatives in 1976, 1978, 1980, and 1982 and the U.S. Senate in 1984 and 1990. He was inaugurated as the forty-fifth vice president of the United States on January 20, 1993, and served eight years as a key member of the administration's economic team, as President of the Senate, a Cabinet member, a member of the National Security Council, and as leader of a wide range of Administration initiatives. He is the author of the #1 bestsellers *The Assault on Reason* and *An Inconvenient Truth*, and of an Oscar-winning documentary movie. His 1992 international bestseller, *Earth in the Balance*, has just been reissued.

Gore and his wife, Tipper, live in Nashville, Tennessee. They have four children and three grandchildren.

JEROME GROOPMAN holds the Dina and Raphael Recanati Chair of Medicine at the Harvard Medical School and is Chief of Experimental Medicine at the Beth Israel Deaconess Medical Center. He serves on many scientific editorial boards and has published more than one hundred fifty scientific articles. His research has focused on the basic mechanisms of cancer and AIDS and has led to the development of successful therapies. His basic laboratory research involves understanding how blood cells grow and communicate ("signal transduction"), and how viruses cause immune deficiency

and cancer. Dr. Groopman also has established a large and innova-
tive program in clinical research and clinical care at the Beth Israel
Deaconess Medical Center. In 2000, he was elected to the Institute of
Medicine of the National Academy of Sciences. He has authored several editorials on policy issues in *The New Republic*, the *Washington
Post*, the *Wall Street Journal*, and the *New York Times*. He has published four books; his most recent, the bestseller *How Doctors Think*,
came out in a paperback edition in March, 2008.

"I became interested in the subject," he says, "because several psychiatrists and psychologists told me that they were deluged with referrals of bipolar disease in children and worried that it had become
a diagnosis du jour. Following its appearance, there was a robust
debate in the popular press as well as on several television programs
focusing on the controversy. The death of the toddler in Massachusetts who was medicated since the age of twenty-eight months for
bipolar disease with drugs that have not passed FDA approval for
this indication has sparked ongoing litigation. Researchers and clinicians will hopefully arrive at consensus criteria for diagnosis and
have better data about the risks and benefits of different therapies."

STEPHEN S. HALL writes about molecular biology, genetics, and
the intersection of business and biomedicine. In addition to author-
ing five books, including *Invisible Frontiers* (on the birth of biotechnology) and *Merchants of Immortality* (on the science and politics
of stem cell research), he writes for many magazines, including the
New York Times Magazine (where he has worked as an editor), *National Geographic*, *The Atlantic Monthly*, *The New Yorker*, *Discover*,
Technology Review, and *Science*, among others. He has also taught
science writing at the Graduate School of Journalism at Columbia
University. He lives in Brooklyn, New York, with his wife and two
children.

He writes, "I brought no particular expertise to the subject of
wisdom—except, like just about everyone else, wishing I had more
of it—when the editors of the *New York Times Magazine* asked me to

write an article about 'wisdom research.' I was more than a little dubious about the premise but soon found myself exhilarated by the intellectual challenge of defining such an elusive human virtue. It turns out that many traditional values associated with wisdom are being explored by modern psychological and neuroscience research. So many, in fact, that now I'm writing a book about it."

AMY HARMON is a National Correspondent for the *New York Times*. In 2008, Ms. Harmon won a Pulitzer Prize for her series "The DNA Age," which examines the impact of genetic technology on American life. In 2001, she shared a Pulitzer for a series on race relations in America. She has received several other awards. Before joining the *Times* in 1997 as a technology correspondent, Ms. Harmon was a reporter at the *Los Angeles Times*. She received a BA degree in American Studies from the University of Michigan in 1990. Ms. Harmon lives with her family in New York City.

"Maybe it is the trick of a mind not designed to grasp a tragic fate foretold in DNA," she muses. "Or maybe it is Katie Moser: her humor, her fundraisers, her text messages. But although our relationship is defined by her genetic destiny—it is, after all, why I wanted to write about her—I often find that I have forgotten that Katie is destined to develop Huntington's disease. When it hits me, again, I am filled anew with powerlessness and anger and urgency. All I can think is, there needs to be a cure."

GARDINER HARRIS is a reporter at the *New York Times*'s Washington bureau, where he covers public health. He joined the *Times* in April 2003. Previously, he was a reporter for the *Wall Street Journal*, the Louisville *Courier-Journal*, and the San Luis Obispo, California, *Tribune*. He is a recipient of the George Polk and the Worth Bingham journalism awards. He is a Yale graduate. JANET ROBERTS is a *New York Times* reporter who specializes in analyzing data and mining documents for the material that informs stories. She has been with the *Times* since 2005. Before that, she was the computer-assisted reporting editor at the *St. Paul Pioneer Press* in Minnesota. BENEDICT CAREY has written for the medical journal *Hippocrates* and the *Los Angeles Times*. Since 2004, he has been a science and medical writer for the *New York Times*.

Speaking for the team, Gardiner Harris writes, "We launched the Minnesota project after discovering a trove of dust-covered records stashed in the file cabinets of the Minnesota Board of Pharmacy. Unfortunately, the records weren't organized, so we spent months getting them entered into a database and cleaning them up. Then we spent more months finding the anecdotes that would tell the stories—a long and difficult process.

"The stories offer the first window into the substantial sums that drug makers provide to clinicians around the country, and the effects this money has on routine care. In the wake of the stories, legislation has been proposed in Congress to create a national registry of these payments, and investigations have been launched that reveal just how poorly such payments are policed by universities."

JOSEPH KAHN is the deputy foreign editor of the *New York Times*. Prior to the *Times*, in 1998, he worked at the *Wall Street Journal*. From 2003 to 2007, he was the *Times*'s Beijing bureau chief; in 2006, he and Jim Yardley shared a Pulitzer Prize for International Reporting.

BEN MCGRATH has been a staff writer at *The New Yorker* since 2003. He has also written for the *New York Times, Slate*, and the *New York Observer*, among other publications.

"When I began reporting this story," he recalls, "there was some buzz surrounding the planned revival, on NBC, of the old nineteen-seventies hit *The Bionic Woman*. Claudia Mitchell's continued progress—she now has a sense of touch in her prosthesis—seems to undermine the fantastical premise of such a show, which may explain why it has already come and gone. Why use science fiction to portray what science has already begun achieving?"

TARA PARKER-POPE writes the popular blog Well for NYTimes .com as well as the weekly Well column for the paper's "Science Times" section. Before joining the *Times* in August 2007, Tara worked for the *Wall Street Journal* for fourteen years, working as the paper's weekly health columnist and as a reporter in Dallas, London, and New York. She began her career at the *Austin American-Statesman* and the *Houston Chronicle* and is a graduate of the University of Texas at Austin. She has written two books: *Cigarettes: Anatomy of an Industry from Seed to Smoke* and *The Hormone Decision*, based on an award-winning article about the country's largest clinical trial of women's health. She is working on her third book, *The Science of Marriage*, to be published by Dutton in 2009.

RICHARD PRESTON's critically and commercially acclaimed books have cemented his status as a leading nonfiction journalist and gifted storyteller. He is the bestselling author of eight books, including *The Hot Zone* and *The Wild Trees*, and most recently *Panic in Level 4*. He is a regular contributor to *The New Yorker* and has won numerous awards for his writing. He is the only nonmedical doctor to receive the Champion of Prevention Award from the Centers for Disease Control. As a result of his contributions, an asteroid has been named "Preston." Preston is a lump of rock three miles across, which could one day collide with Mars or the Earth, causing an explosion similar to the one that is believed to have wiped out the dinosaurs.

Preston lives near New York City with his wife and three children. He is at work on a children's book about climbing giant redwood trees. He can be found on the web at www.richardpreston.net.

Writing this article was a saga, he notes: " 'An Error in the Code' took me seven years to write. (*New Yorker* pieces always seem to take a long time, but this was ridiculous.)

"It describes a genetic disease, called the Lesch-Nyhan syndrome, caused by a tiny error in a person's genetic code, which gives the person an irresistible compulsion to, in effect, cannibalize himself. The person bites off his extremities and portions of his own face, and otherwise attacks himself, as well as attacking the people he loves. The disease is caused, typically, by the alteration of only a single letter in the more than three billion letters of the human DNA.

"The disease made its victims seem inhuman. I couldn't find a way into writing the story, despite spending months and finally years on it. I needed to understand, as a writer, what it might feel like to *have* the disease, to connect the seemingly unknowable experience of self-cannibalism with that of common humanity. The only way to find humanity in the story was to become friends with two men who had the disease, to ask them to describe their experience to me, and ultimately to look through the window of their minds into the human condition we all share."

TINA ROSENBERG writes for the *New York Times Magazine*. Until 2007, she was an editorial writer for the *Times*. She has written two books, *Children of Cain: Violence and the Violent in Latin America*, published in 1990, and *The Haunted Land: Facing Europe's Ghosts After Communism*, published in 1994. The latter book won the Pulitzer Prize for General Non-Fiction and the National Book Award. She has had a long career as a freelance writer for magazines, including *The New Yorker*, *The Atlantic Monthly*, *Rolling Stone*, and *Foreign Policy*, specializing in Latin America, globalization, human rights, and public health. She has lived in Nicaragua, Chile, and Mexico, and lives in New York City with her family.

"This article was a departure for me," she explains, "since I don't usually write about domestic issues. What most surprised me when I started researching the issue was the extent of the misconceptions I held about treating pain and addiction—and that the vast majority of Americans, even doctors, even police officers who investigate these kinds of cases, hold the same erroneous beliefs. I did not know, for example, that there is no necessary ceiling on the amount of opioid drugs that can be used safely, nor that addiction to them is pretty rare. One of my characters—a pain doctor who is also a pain patient and uses a wheelchair—asks, in the story, why doctors are not trained about pain and how to treat it. I would also like to understand why there hasn't been more of a citizens' movement around this issue. In an aging population, the number of Americans living with chronic pain is steadily rising. My hope in writing the article was that I would open people's eyes to a hugely important issue that, until now, has been largely ignored or misunderstood."

OLIVER SACKS is a professor of neurology and psychiatry at Columbia University and was recently appointed Columbia's first University Artist. He is a frequent contributor to *The New Yorker* and *The New York Review of Books*, and is best known for his neurological case studies, which are published in books such as *The Man Who Mistook His Wife for a Hat* and *Awakenings*. His essay on Clive Wearing is included in his latest book, *Musicophilia: Tales of Music and the Brain*. He is working on a book about vision and the brain. For more information, see www.oliversacks.com.

"I first learned of Clive Wearing through Jonathan Miller's film in 1986," he writes, "and his story remained in my mind, stimulating a great deal of thought, for many years. But it was not until 2004, when I read the proofs of Deborah's own extraordinary book about her husband, that I resolved to visit them and see his situation for myself. The day after I returned home, I sat down and wrote 12,000 words of 'notes.' Some of my essays are revised extensively, and others are published essentially as they were first written. Clive's story is one of the latter; there have been additions, but no essential changes to the piece as I originally wrote it."

SALLY SATEL is a resident scholar at the American Enterprise Institute and the staff psychiatrist at the Oasis Clinic in Washington, D.C. Dr. Satel was an assistant professor of psychiatry at Yale University from 1988 to 1993 and is now a lecturer. She has written widely in academic journals on topics in psychiatry and medicine, and has published articles on cultural aspects of medicine and science in numerous magazines and journals. Dr. Satel is author of *PC, M.D.: How Political Correctness Is Corrupting Medicine* (Basic Books, 2001) and the coauthor of *One Nation Under Therapy* (St. Martin's Press, 2005).

"My interest in the organ shortage is organic—that is, it grew out of my personal experience as a kidney transplant recipient in 2006," she explains. "My glorious friend, Virginia Postrel, gave me her kidney but until she came along I was worried I'd spend over five years on dialysis waiting on the national transplant list for a cadaver organ. That is the fate of over 75,000 people today. I am committed to changing the law so that government can compensate people for relinquishing a kidney. This will save thousands of lives a year."

JOHN SEABROOK is a staff writer at *The New Yorker*. His books include *Deeper: My Two-Year Odyssey in Cyberspace*, *Nobrow: The Culture of Marketing, the Marketing of Culture*, and the forthcoming *Flash of Genius: and Other True Stories of Invention*, due in September, 2008. A film based on the title story will be released in October, 2008. Seabrook lives with his wife and son in New York City.

"Like Cary Fowler, the main character in 'Sowing for Apocalypse,' I grew up on a farm," he says, "and like him I found that my intellectual interests made farm living too constrictive. Throughout my career I have tried to combine my formative agricultural experience with my literary pursuits, by writing about food, plants, and farmers. The Seed Vault was an ideal subject for me."

MARGARET TALBOT is a staff writer for *The New Yorker* magazine, where she often writes on science and the law, and a senior fellow at the New America Foundation. She lives in Washington, D.C., with her husband, writer Arthur Allen, and their two children.

"One of the things that continues to amaze me about the whole area of lie detection," she notes, "is how many people volunteer to take polygraphs every day in all kinds of circumstances, and how alluring the idea of an even more powerful truth machine continues to be—no matter how much valid skepticism we air about the enterprise."

JIM YARDLEY is Beijing Bureau Chief for the *New York Times*. He has been a reporter at the *Times* for more than ten years, including the last four in Beijing. He has written about a wide range of topics in China, including environmental problems, inequality, peasant protests, migration, and the pressures of rapid development. He has won several awards for his work in China, including the 2006 Pulitzer Prize for International Reporting, which he shared with his colleague, Joseph Kahn.

"I have written several stories about water problems in China," he says, "whether about water scarcity, water pollution, or even the water politics of dam building. In this one, I tried to capture the startling paradox of how officials keep pushing wild development even as water is steadily disappearing beneath the North China Plain. Sometimes when I drive through the region, and I see barren riverbeds and bridges built to span water that no longer exists, I wonder how the problems can be reversed. Of all the challenges facing China, none are more significant over the long haul than the country's lack of available, clean water."

CARL ZIMMER is the author of six books, including *Evolution: The Triumph of An Idea, Parasite Rex*, and, most recently, *Microcosm: E. coli and the New Science of Life*. He began to write about science at *Discover*, where he served as a senior editor from 1994 to 1999. He writes frequently for the *New York Times*, as well as for magazines such as *Scientific American, National Geographic*, and *Time*. In 2007, he was awarded the National Academies Communication Award "for his diverse and consistently interesting coverage of evolution and unexpected biology." His Web site is carlzimmer.com.

"I've been fascinated for years by the way evolution produces new diseases," he writes. "But I learned mostly about diseases caused by viruses and other parasites—things that could experience natural selection. Cancer seemed different, just a case of our bodies going haywire. So it was a huge surprise to discover that cancer is an inevitable by-product of our evolution as animals, as collectives of trillions of cells. The ideas I write about in the article are very new—so new, in fact, that one of the scientists I interviewed essentially broke off our interview so he could start planning an experiment on cancer biology that occurred to him during our conversation."

Permissions

A Note from the Series Editor

Submissions for next year's volume can be sent to:

Jesse Cohen
c/o Editor
The Best American Science Writing 2009
HarperCollins Publishers
10 E. 53rd St.
New York, NY 10022

Please include a brief cover letter; manuscripts will not be returned. Submissions can be made electronically and sent to jesseicohen@ netscape.net.

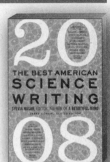
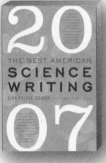